Cultures
Beyond
the Earth

CULTURES BEYOND THE EARTH

Edited by Magoroh Maruyama
and Arthur Harkins

VINTAGE BOOKS

A Division of Random House, New York

First Edition A VINTAGE ORIGINAL, August 1975
Copyright © 1975 by Magoroh Maruyama and Arthur Harkins
Foreword Copyright © 1975 by Alvin Toffler

Library of Congress Cataloging in Publication Data
Main entry under title:
Cultures beyond the earth.
    The 1974 meeting of a symposium held at the annual
meeting of the American Anthropological Association;
publications of the 1971 symposium are entered under
American Anthropological Association Experimental Sym-
posium on Cultural Futurology.
        1. Extraterrestrial anthropology—Congresses.
I. Maruyama, Magoroh.   II. Harkins, Arthur M.
III. American Anthropological Association.
GN27.C85   301.2   75–12681
ISBN 0–394–71602–7

Manufactured in the United States of America

Contents

Foreword: The Cosmic Rebound

ALVIN TOFFLER

> . . . I love my poor earth
> because I have seen no other
> —Osip Mandelstam

If life in outer space does not exist, we are justified
in inventing it. All individuals are influenced by the
glancing images that rebound from the mirrors
around them. We shape our personae in response to
our own reflections in the social looking glass. In
quite the same way, whole cultures in today's densely
inter-communicative world are affected by the reflections they produce in other cultures—real or imagined. To put it another way: What we think, imagine or dream about cultures beyond the earth not
only reflects our own hidden fears and wishes, but
alters them. That is why this short, closely packed,
surprising little book is so important. It forces us to
disinter deeply buried premises about ourselves.

Some years ago a group of university students and

I sat around a classroom discussing the links between values, technology, and the future. The talk was meandering, vague, unsatisfactory. No one, it seemed to me, was uttering anything remotely brilliant, or even worth remembering. Nevertheless, one young woman continued energetically taking notes throughout. She wrote and wrote desperately trying to keep up. Poor girl, I thought. Can't tell what's important and what isn't. Finally my curiosity won out. I asked her what she was writing.

"I'm keeping a list of all the assumptions I hear in the conversation," she replied coolly.

She took my breath away, and a dead, impressed silence fell over the classroom. She was, of course, behaving in an admirably intelligent fashion, operating at a level above the rest of us—a "meta-level," as it were.

For more than a century anthropologists who study so-called primitive cultures have held up a reflecting mirror to the assumptions of industrial society. In this glass we are able to recognize our own ethnocentricism, our narrowly materialist values, our powerful yet limiting assumptions about time, space, logic and causality. But other aspects of our own culture remain unreflected. Because the cultures examined by anthropology have been, for the most part, less technologically advanced than our own, less differentiated and less rapidly changing, vast reaches of our own way of life are unilluminated by either contrast or comparison. It is as though the light reflected backward on ourselves left large patches of dark shadow.

The darkness of these patches is intensified by the very fact that the other cultures under study by anthropology have been *human* cultures like our own, meaning that we share with them not only common body forms and sensory apparatus, but common needs for food and energy, a common capacity for verbal expression and common reproductive systems—all of which subtly structure our assumptions about reality. In contrasting ourselves with other humans we can only go so far.

It is here that "extraterrestrial" anthropology, the subject of *Cultures Beyond the Earth*, comes into play. Precisely because it deals with cultures that are (or, more exactly, might be) more technologically advanced than our own, as well as less so, because it deals with life-forms radically different from our own, it ultimately casts light on some of the hidden reaches of our own culture. It raises the critique of our cultural assumptions to a "meta-level."

Thus it calls into question the very idea of cultures based on a single epistemology, of single time tracks or merely human sensory modalities. It forces questions about intelligence and consciousness. It makes one wonder whether our assumptions about probability apply universally. In the course of all this, it also begins to give intellectual shape to the whole question of space exploration and its relationship to our world.

It examines the reasons for space colonization, the various possible means by which we might make contact with extraterrestrial life, the types of contact, the possible responses, the obstacles to cross-cultural communication, the nature of the human skills required in establishing space settlements, and a variety of other equally unusual and fascinating issues.

It argues that when we begin actually planning extraterrestrial bases (and even as I write this, a conference is under way at Princeton on the possibility

and nature of manufacturing operations to be conducted in outer space), we had better see to it that
the colonies we plant are culturally diverse, that they
are not simple, stripped-down versions of Western
industrialism. Yet so deep is our own cultural conditioning that even the authors of *Cultures Beyond the
Earth* are frequently trapped by it. Thus in the
fictional scenarios which form part of the book, they
picture human moon colonists celebrating Hanukkah and Christmas jointly, in an apparent excess of
ecumenicism; but no one celebrates Ramadan or the
Chinese New Year. Several times we find the assumption that space colonies will have to be technocratic,
almost military in structure because of harsh environments—as though totalitarian rule is necessarily
"efficient." A number of writers suggest that the
first human colonies will be launched by a "united"
human race in anticipation of, or in response to,
earthside disaster—a familiar science-fiction assumption that ought, perhaps, be sent into retirement.
Such lapses, however, are rare in *Cultures Beyond
the Earth,* and they merely underscore the difficulty
of this kind of cultural analysis and criticism. What
is important is not the book's unevenness (all collec-
tions are uneven) but its striking originality. No
doubt it will be attacked by pedants and purists.
But it springs from so innovative an idea and is
carried out with such intellectual *brio* that it should
be read by anyone seriously interested in anthropology, space, or both.

At a time when our industrial culture is disinte-
grating and we face the task of designing a wholly
new civilization, we need to look at ourselves through
new mirrors, in new ways. *Cultures Beyond the Earth,*
one might say, sends a signal into the depths of outer

space which, in turn, bounces back toward us and showers us with insights. It makes us all the fortunate beneficiaries of what might best be called "cosmic rebound."

London

Contributors

Roger Williams Wescott, born in Philadelphia in 1925, graduated summa cum laude from Princeton in 1945. After receiving his M.A. in Indology and his Ph.D. in Linguistics at Princeton, he held a Rhodes Scholarship at Oxford. He taught history and human relations at MIT and Boston University, and English and Social Science at Michigan State, where he was also Director of the African Language Program. The author of thirteen books, including *Divine Animal,* he is now Professor of Linguistics and Chairman of the Anthropology Department at Drew University in New Jersey.

Donald K. Stern is a student at the University of Washington, majoring in extraterrestrial sociology.

He has done research on kinship and cloning, neo-medievalism in the Northwest, and the role of theology in science fiction.

Barbra D. Moskowitz is a graduate student in anthropology at SUNY Binghamton and is working on an archeological novel set in pre-Columbian Southwestern America. She obtained her B.A. in Creative Writing at State University of New York and worked at the Interscience Division of John Wiley and Sons.

Philip Singer obtained his B.A. at Syracuse University, studied at the Universities of Oslo and Delhi, and received his Ph.D. in anthropology from Syracuse. He has been on the faculty of Syracuse University, SUNY Albany, Albany Medical College and California State College, Los Angeles, and is currently Professor of Anthropology at Oakland University in Michigan.

Carl R. Vann received his A.B., M.A. and Ph.D. from Syracuse University. He has taught political science and community medicine at Syracuse, Wayne State, Oakland and Michigan State, and is currently chairman of Allport College of Human Behavior at Oakland University and Professor of Community Medicine at Michigan State.

Bill Gerken, Jr., is a free-lance writer and consultant on alternative futures. He received his M.E. from Stevens Institute of Technology and his M.B.A. from Florida State. He worked as a rocket test engineer and field engineering representative for Bell Aerosystems Company and as a Senior Flight Systems Engineer for ILC Industries, Inc., at the JFK Space Center in Florida. Earlier he was the ILC Flight Team Engineer in charge of spacesuit preparation for the Apollo Twelve Lunar Landing Mission.

Shirley Ann Varughese was born in Ohio in 1950 and is now living in Staten Island, New York. She is interested in alternative life styles.

Mary Elizabeth Oberthur studied at Indianapolis Methodist Hospital School of Nursing and Indiana University, and is a registered nurse. She has been a psychiatric and neurosurgical nurse and Chairman of the Nursing Education Program Committee at Indianapolis Methodist Hospital.

Kim Arthur Mayyasi wrote his chapter as a senior at Deerfield Academy. He was a regional winner of a NASA-sponsored Skylab Student Project and built a high-performance hovercraft. Now a student at MIT, he hopes to major in electrical engineering with interests ranging from micro-electronics to the possibilities of communication with extraterrestrial intelligence.

Cultures Beyond the Earth

Introduction

MAGOROH MARUYAMA

The contents of this book may come as a surprise to many readers because its direction of inquiry is different from what is usually found in books of science or science fiction. This book is a beginning effort to free us from our traditional Western notions of society, culture, life style and logic, and earthlings' definition of civilization, intelligence and emotion. This book endeavors to infuse in our technological thinking some nontraditional concepts of society learned from the life styles and logics in various cultures existing on the earth in order that we may incorporate in our design of extraterrestrial human communities a wider range of possible social and cultural forms. At the same time, we cannot exclude the possibility of contact with nonhuman forms of civilization, intelligence and psychology, and we need to prepare ourselves in the method of understanding and communicating with them. The methods developed by anthropologists in the study of human cultures different from our own can be extended to include nonhuman cultures, even though some of the nonhuman cultures may be more advanced than our own and may be beyond our comprehension.

There is a joke about what two men and a woman, shipwrecked on an island, will do:

- If they are *Spanish,* one of the men will kill the other.
- If they are *Italians,* the woman will kill one of the men.
- If they are *Koreans,* the woman will kill both men.
- If they are *Japanese,* one of the men will feel the awkwardness of the situation, and in order to make the situation easier for the others he will commit suicide. Then the second man will blame himself for the death of the first man and will commit suicide. Then the woman, feeling responsible for the death of the two men, will commit suicide.

If there are so many culturally different "solutions" to the same situation involving only three persons, how many solutions would there be if we were to design an extraterrestrial community containing twenty, fifty, two thousand or ten thousand individuals?

In planning such a community, we would make decisions not only on the design of the physical environment, but also on the family structure, the communication networks and the political system. We would not be restricted to the one-father, one-mother system, bipartisan government or majority rule. We would explore many other sociocultural alternatives and the philosophical assumptions underlying them before making our choices. Not to do so would be uneconomical and could deprive us of the opportunity to enrich human life.

This book is the result of a series of events. In 1970 I organized a symposium on cultural futuristics as part of the American Anthropological Association annual meeting. The symposium became an annual program, and Professor Harkins and I chaired it alternately. During the first years the symposium was organized around invited papers. But we felt a need for fresh ideas, and in 1973 we changed our format and based the symposium on an essay/fiction contest on cultural alternatives which we held only among anthropologists. For the 1974 symposium we chose the topic of extraterrestrial communities and expanded the contest to include technologists and freelance writers. Of the eight winning papers, two were written by anthropologists, one by a technologist, two by science students, and three by free-lance writers. This book consists of these winning papers.

The book is divided into two parts. The first part consists of essays dealing with the scope of problems expected to arise in extraterrestrial communities. The second part is composed of sociocultural fiction.

This book challenges the pattern of thinking in which we all, including scientists, technologists, engineers, policy makers, and even social scientists often get trapped. It also attempts to remedy the ethnocentrism so prevalent in European and North American humanism, ideologies and visions of the future. More specifically, it challenges the homogenistic, universalistic philosophy—the belief in one truth and one logic—which is erroneously considered to be the basis of "scientific" thinking but which is now proving to be unscientific.

Since this book attempts to introduce different logics into technological thinking, and since many physical scientists and social scientists may be unaware that there *are* different logics, let me briefly discuss what is meant by "logical" or "scientific."[1]

The physical and biological sciences have progressed from theories of the never-changing universe, the thermodynamically decaying universe, Einstein's universe of relativity, the probabilistic universe of quantum mechanics to the more recent theory of a mutual causal network which can explain increase of differentiation, complexity, heterogeneity, symbiotization and growth in the universe. In this more recent theory, many elements of a system can cause one another, as compared to the older theories which ruled out mutual causation as illogical. There are several types of mutual causation. One of them is the deviation-counteracting type, such as is found in the ecological balance between the number of predators and the number of prey, or in the physiological system that maintains constant body temperature. Another type amplifies differences and generates heterogeneity. This type is responsible for the evolutionary process, the growth of an embryo into the adult structure, the growth of cities and many other biological and social processes of development and self-organization. An example is the evolution of one species of moth and the species of bird which eats that moth. The mutants of the moth species that are more camouflaged than the average will survive better, and the mutants of the bird species that are more clever than the average at discovering the camouflaged moth will survive better. Consequently the moth gets more and more camouflaged, and generation after generation the bird develops more and more ability to discover the camouflaged moth. Both the moth and the bird become more and more complex. Or consider a homogeneous grassland. A pioneer comes and settles at one place. The choice of the place may be accidental: his horse gets sick, and he can go no farther. But once he settles, other pioneers may come to the same place. Farms develop and shops open because of the mutual amplification of activities between the settlers. A city grows, and the grassland is no longer homogeneous.

Even some of the nonbiological processes show characteristics of differentiation-amplifying mutual causation. Random noises may organize themselves into nonrandom patterns by means of resonance, and we can speak of self-organizing noises. Evenly distributed gaseous matter in an astronomical space may condense into heavy bodies due to mutual gravity between the molecules, making the space less homogeneous.

The universe based on this theory of mutual causation in biological, social and physical processes is a self-organizing one, quite different from the never-changing universe of the Middle Ages or the decaying universe of thermodynamics.

Newtonian celestial mechanics coincided with the historical period during which a philosophy based on the concepts of substance, uniformity, permanence and stability dominated Western thinking. The belief in one truth and one logic culminated in this period, and homogeneity was assumed to be the basis of the universe. A few centuries later, the theory of thermodynamics formulated the idea that heat moves from the warmer parts of a system to the cooler parts, and therefore that heat becomes more and more evenly distributed in an isolated system. This gave rise to the philosophy that the universe changed from heterogeneity to homogeneity and reinforced the earlier homogenistic philosophy on the basis of a different principle. It amounted to a philosophy of a decaying universe, and it could not explain the biological and social processes that generate heterogeneity and complexity. It simply begged the question by saying that a living organism is not an isolated system.

It was not until the 1960's and 1970's that the *logical* difference between the thermodynamic processes and the biological and social processes became clarified: the thermodynamic theory was based on the assumption that *random and independent* occurrences of events (like tossing a coin) are the basis of the universe, while the biological and social processes occur by means of a *mutual* causal network. If the events are random and independent, the most probable states in which a system finds itself are states of even distribution (heads and tails without a definite pattern). It follows that if a system is in an improbable state of uneven distribution, it will tend to change to a state of more even distribution. Hence the universe changes from a heterogeneous state to a more homogeneous state. On the other hand, a *mutual* causal network can generate patterns, heterogeneity and complexity, as we have seen.

The failure of earlier theories to recognize and formulate mutual causation and its various types was due to the fact that classical Greek logic prohibited "circular" reasoning, even though mathematics could handle mutual interactions. What is considered "scientific" depends on the type of logic one uses. The recent recognition of mutual causal processes in science marks a transition from Greek to non-Greek logic. There are many non-Greek logics that correspond to the theory of a heterogeneous, self-generating universe of mutual causation. For example the logic of the Mandenka in Western Africa[2] and Navajo logic[3] are heterogenistic, mutualistic, relational, contextual, harmonistic and symbiotic as compared to the homogenistic, hierarchical and classificational Greek logic.

Most Americans have been brought up with the homogenistic logic, and therefore it is only "logical" for them to assume that there must be a universal formula that works for everybody, and that what works for Americans should work for anybody else. This is the logical basis of the ethnocentrism found among many American scientists, technologists, engineers and policy makers.

When introduced into different cultures or different logics, most Westerners ask two questions: (1) Don't the differences create conflicts? (2) If there are differences, who is right and who is wrong? These questions are themselves trapped in Western logic.

Western logic considers heterogeneity a source of conflict. But consider the following: Animals convert oxygen into carbon dioxide, and plants convert carbon dioxide into oxygen. They do exactly the opposite, yet they do not conflict. On the contrary, they help and need one another. Heterogeneity also enables diversification of resource requirements. If all animals were to eat the same food, there would be a food shortage and conflict. The richness of life in tropical rain forests or on coral reefs is the result of the heterogeneity of the species. Heterogeneity also provides a higher survival probability in case of change of climate or other catastrophes. The Mandenka say that homogeneity, rather than heterogeneity, is the source of conflict. They say that if people are forced to be similar, the only way they can be different is to create a vertical hierarchy and try to subordinate one another, and this creates conflicts. Westerners talk about "unity by similarity," while non-Westerners live in "harmony thanks to diversity." These two ways of thinking are based on two entirely different logics.

Westerners are also concerned about determining who is right and who is wrong when there are differences. But consider the following: Your two eyes give you two slightly different images of the same object. Do you ask which image is right and which is wrong? The difference between the two images is not disturbing; on the contrary, it is useful. Each image is two-dimensional. But the differential between the two images enables the brain to compute the third, invisible dimension. While Westerners insist on *agreement,* many non-Westerners benefit from cross-subjectivity and poly-ocular vision.

The psychological need to seek and depend on *one* truth, *one* theory or *one* right answer seems to be related to the Western family structure of one father, one mother and their children. In this system the child relates mainly to one pair of adults, depends on one authority and learns one way of thinking and doing things. In many other cultures the child relates to several adults who may be relatives or non-relatives, has diversified emotional sources and learns different points of view and ways of doing things. In some cultures children are exchanged periodically between families. In these cultures there is less tendency to rely on one truth. Future generations of humans, whether living on the earth or beyond the earth, may not have the same family system as ours, and may therefore be less ethno- and anthropocentric than ourselves.

This book may come as a culture shock because it pursues an unexpected direction of thought. Rather than provide specific answers, it attempts to raise questions. The more different kinds of questions the book raises and the more thought and discussion it stimulates, the more rewarded the authors and the editors will feel.

Notes

1. A fuller discussion of what is to follow is found in M. Maruyama, "The Second Cybernetics: Deviation-Amplifying Mutual Causal Processes," *American Scientist* 51 (1963) : 164–179, 250–256; and M. Maruyama, "Paradigmatology and Its Application to Cross-Disciplinary, Cross-Professional and Cross-Cultural Communication," *Cybernetica* 17 (1974) : 136–156; 237–281.

2. Sory Camara, "The Concept of Heterogeneity and Change among the Mandenka," in *Cultures of the Future,* ed. M. Maruyama and A. Harkins (The Hague: Mouton Publishers, 1976).

3. Clyde Kluckhohn, "The Philosophy of the Navaho Indians," in *Ideological Differences and World Order,* ed. F. S. C. Northrop (New Haven, Conn.: Yale University Press, 1949) .

1.
Toward an Extraterrestrial Anthropology

ROGER W. WESCOTT

Man's interest in the extraterrestrial is at least as old as the Upper Paleolithic, when early Europeans incised lunar markings on ivory proto-calendars. Deities from nearly every culture have been identified with one or another of the closer solar planets. And systematic astronomical investigations and astrological beliefs have been a part of the intellectual life of every known literate society.

Since the dawn of the Space Age in the 1950's, most of the physical sciences have been adjoined to astronomy in the quest for effective interdisciplinary collaboration in the exploration of the circumterrestrial and lunar environments. Even the life sciences have been partially coöpted in this quest, although more skeptical biologists like George Gaylord Simpson have warned self-styled "exo-biologists," or would-be students of extraterrestrial life, that unless they can soon demonstrate the palpable existence of their chosen subject matter, they may shortly come to be regarded as pursuers of phantasms and accordingly stigmatized as "ex-biologists."

As one of the youngest sciences, anthropology has been relatively uninvolved thus far in extraterrestrial pursuits. Being by definition an anthropocentric discipline, it has remained predictably geocentric overall. The only substantial exception to this generalization is the predominantly anthropological subdiscipline of folklore, with its substantial component of celestial mythology.

If, as I believe, anthropology is to turn its focus in an increasingly extraterrestrial direction, it is worth asking what strengths and weaknesses anthropology brings to extraterrestrial tasks. Anthropology's most salient strength is probably its range, which embraces the entirety of earth's land surface, the duration of the Pleistocene and Holocene epochs (now estimated as three million years), and the cultural inventory of all human societies, primitive as well as literate and extant as well as extinct.

Anthropology's weaknesses, however, are almost as conspicuous as her strengths so far as extraterrestrial competence is concerned. Largely as the result of an informal division of intellectual labor with sociology, whereby sociology has been granted primary jurisdiction over Western urban civilization, anthropology has focused most of its attention on preurban and non-Western cultures. The result of this focus has been an anthropological bias in favor of the primitive and the prehistoric. In view of the fact that all recent outer-space enterprises have been launched by the industrialized nations, either Western or Westernized, this anthropological bias has undoubtedly worked to the disadvantage of anthropology as regards its adaptation to man's new interplanetary thrust.

On balance, anthropology probably will be successful in adjusting itself to a less parochial planetary orientation, largely because of its holistic approach to knowledge and its methodological flexibility.

But what the results will be of such a redirection of anthropological concern remains a question. Just as exo-biologists now run the risk of being called ex-biologists, so may anthropologists with extraterrestrial interests find themselves regarded with suspicion by the more conservative members of their own profession. Even if this happens, however, adventurous anthropologists may console themselves with the knowledge that extraterrestrial studies are intrinsically mind-stretching and that they are doing more than their more routinized colleagues to stimulate the too often sluggish anthropological imagination. Moreover, the chances are that, for every colleague who threatens extraterrestrializers with academic ostracism, there will be another who lauds their intellectual enterprise. In short, there is no reason why those anthropologists who are interested in studying the emergent subject of man in space should hesitate to give free rein to that interest.

Extraterrestrial anthropology need not, of course, be restricted to the study of the psychological and social responses of the small astronautic teams to which our space effort is currently confined. Indeed, its ethnological component can hardly be said to be functioning at all until such time as ethnologists begin to observe and describe the nascent cultures and subcultures of human communities in earth orbit, in lunar orbit, or on the lunar surface. And it will be surprising if these communities are not eventually extended beyond the earth-moon system to other planets and their satellites, both natural and artificial.

Because, however, of the obvious possibility that in the course of our interplanetary explorations, we may encounter intelligent beings of other species, extraterrestrially oriented anthropologists must continue to cultivate their biological as well as their ethnological competences—not a difficult endeavor, because there are many well-known humanoid (in the strict sense of the term) species here on earth with which they can better familiarize themselves. (Typologically speaking, any species is humanoid if either its morphology or its behavior is conspicuously analogous to our own.) Behaviorally humanoid taxa range from social insects through cephalopods and corvid birds to cetaceans and elephants. And the great apes are morphologically as well as behaviorally humanoid. We know less about some of these animals—especially the giant squid and the killer whale—than we would like to. But ethnological literature on them is copious, and some of it, gratifyingly, is of anthropological authorship, as in the case of Gregory Bateson's work with octopuses[1] or of Burt and Ethel Aginsky's work with dolphins.[2]

Thus far, we have been using the term "extraterrestrial" in its spatial sense, to refer to whatever is external to the earth and its atmosphere. Yet there is a nonspatial sense in which "extraterrestrial" is synonymous with the colloquial term "unearthly" and refers to whatever is so uncommonly experienced as to put its reality in doubt. Among earthbound "extraterrestrials," those which have most interested anthropologists in recent years are such

controversial hominoids as the apelike Yeti of the
Himalayas and the manlike Sasquatch of the Cas-
cade Mountains, both of which are popularly referred to as "abominable snowmen." Carleton Coon[3]
and Grover Krantz[4] have gone on record as believ-
ing that these elusive creatures merit further search.
And John Napier has written a book about them.[5]

Other humanoids whose existence and planetary
origin are more than a little doubtful are the oft-
reported occupants of "flying saucers" and other
unidentified flitting or floating objects. Even if the
reality of such beings is granted, it is altogether
unclear whether they are reconnoitering visitors
from other planets or fellow natives of earth whose
coterrestriality we have overlooked because for the
most part they keep themselves out of view—perhaps in underwater habitations. In either case, however, they are psychologically unearthly to us.

One way to cope intellectually with such creatures as "flying saucerites" is to regard them as the
mythical beings proper to the Space Age, as angels
were the mythical beings proper to the biblical period. In that case, both orders of beings fall under
the anthropological rubric of folklore and can be
dealt with by ethnopsychologists in terms of symbolic
projection of the major concerns of the era in
question.

On the other hand, such an explanation of widespread belief in the reality of nonhuman but humanoid beings is undeniably reductionistic. If, in
keeping with the professedly holistic objectives of
anthropology, we attempt rather to frame an ex-
planatory hypothesis that enlarges reality instead of
constricting it, we may be pushed into rethinking
not only the principles of our discipline but also the
nature of the universe in which we live. One such

way of rethinking things is to assume that the "real" universe has either more than the three dimensions of space which we perceive or a different type of time flow from the one we conventionally postulate. Time —as Western man has conceived of it at least since the Renaissance—is single in dimension, uniform in pace and irreversible in direction. If time should turn out to have more than one dimension, discontinuity of pace, or reversibility of direction, or if space should turn out to have more than three dimensions, then it would be quite possible for solidly and prosaically material beings from the "real" world to pass through our illusively constricted space-time continuum as a needle passes through a piece of cloth. In that case, we would probably find such beings fantastic, either in the sense that not being "native," or confined to our world, they strike us as incredibly monstrous, or that because they seem to us to materialize and vanish inexplicably, we dismiss them as hallucinations or hoaxes.

Rather than existing in space and/or time in the conventional sense of these terms, our planet may exist in "hyperspace" and/or "hypertime," where hyperspace is understood to mean space with four or more dimensions and hypertime to mean time which permits events and processes to occur in other than an irreversibly linear and unidirectional manner. On such a "hyperhistorical" sphere, or alternatively such a historical "hypersphere," all the supernatural beings and all the miraculous occurrences known to us from religion and folklore would become explicable as intrusions from the larger earth of reality into the smaller earth of our self-habituation.

Because this way of conceiving phenomenality is so unfamiliar both to common sense and to social science, it may be advisable to employ another simile to clarify it. Let us picture ourselves as travelers on a road, which we may acceptably call "the path of humanity." All of us move at exactly the same pace, which explains why those who were ten years older than us when we were children are still precisely ten years older than us when we reach adulthood. All of us wear blinders, which prevent our seeing anything other than our road, such as time not on our track or space not in our ambit. And most importantly and most remarkably, we all walk backwards, with a good view of the road behind us but no view at all of the road ahead of us, which explains why we can remember the past but not the future.

From this point of view, extraterrestrial anthropology comes to mean not only interplanetary anthropology, but also the anthropology of our own planet with additional spatial or temporal dimensions added to it—what might be called hyperplanetary anthropology. For such anthropology, the chief supporting disciplines would not be astronomy and astronautics, but as regards its historical aspect, mythology and folklore, and as regards its synchronic aspect, the emerging field of investigation that I call "anomalistics," the systematic study of anomalies. (See appendix.)

Another auxiliary discipline which, psychologically speaking, has much in common with mythology is oneirology, the study of dreams. Oneirology, which in antiquity was most closely associated with fortunetelling, came in the nineteenth century to be ancillary to psychoanalysis and since World War II has been most closely tied to neurophysiology. For cultural anthropologists, one of the most interesting aspects of dreams in any culture is that the dream life of a given individual is likely to differ more from his own waking life than it does from the waking life of other known cultures, no matter how alien or "exotic" these may be.

For biological anthropologists, on the other hand, what is noteworthy about dreams is that judging by their physiological correlates, such as brain-wave patterns and eye movements, they are shared by man with other warm-blooded vertebrates but, so far as we can tell, with no other organisms. Further, what is striking about birds and mammals in contrast to other animals is that they appear to be the most adaptable of all complex organisms and the most successful in adjusting to rapidly changing conditions. What this fact suggests is that dreaming is a preadaptive device that enables the higher vertebrates to practice coping with unfamiliar conditions.

All of man's arts and sciences may be regarded as elaborations of basic mammalian dream behavior. All of them exercise his imagination and enable him to handle an increasingly wider range of experience. Among these dream elaborations is the literary form we call fiction. Of all types of fiction, the one most congenial to futurists, especially those with extraterrestrial interests, is science fiction. One of the facts noted by futurists is that, particularly in the works of popular science writers like Arthur C. Clarke and Isaac Asimov, science and science fiction seem to be progressively converging. The implication of this observation is that the most imaginative writings of our most knowledgeable thinkers are the ones most likely to help prepare us for life in an unknown future.

Of course, if the bizarre objects and events studied by anomalists are indicative of a larger reality than the one in which we customarily operate, it may be that the type of artistic and scientific imagination which we have been lauding as the

highest form of human creativity is actually not
creativity at all but receptivity. It may be that most
if not all of the works which we have been calling
fiction are in reality only a more or less garbled
version of perceptions that reach us in attenuated
form from other spatial or temporal dimensions. If
so, it follows that there is really no such thing as
fiction and that what we have hitherto regarded as
innovative genius is actually an exceptional sensitiv-
ity to alien worlds and an effective openness to the
messages that are constantly—though in most cases
vainly—reaching us from them.

To return, however, from the hyperdimensional
universe which we only suspect to the dimensional
universe which we know: Before this century is over
man will, I believe, have established permanent
communities on bodies other than this planet. A
plausible order of settlement for such bodies is:

1. the moon
2. Mars
3. Mercury's "twilight" belt
4. a larger asteroid, such as Ceres
5. one of Jupiter's Galilean satellites, such as Gany-
   mede
6. Saturn's largest satellite, Titan (which has an
   atmosphere)
7. Uranus (whose atmosphere creates a "greenhouse
   effect")
8. Neptune's larger satellite, Triton

In addition to the motives commonly suggested for
this colonization, such as intellectual curiosity or the
hope of profitable mining ventures, I suggest at least
one more, no less compelling than the others, and
that is the drive for self-preservation. For if the earth
were to be struck by another large celestial body,
mankind would, in the absence of extraterrestrial

settlements, be catastrophically and irredeemably destroyed. But in Martian or in Jovian lunar communities, not only could human history be preserved
but human evolution could continue even after the
disappearance of the planet that gave birth to our
species.

In addition to such "stationary" communities,
which follow the fixed orbits of the celestial bodies to
which they are attached, we might, as Isaac Asimov
has proposed, develop "migratory" communities of a
size more substantial than that possible in a manufactured spaceship, by occupying some of the smaller
asteroids, hollowing them out and building model
cities inside them. If we were also to equip such
processed habitats with propulsive systems, they
should come to combine the spaciousness of planets
with the maneuverability of spacecraft in such a way
as to allow their inhabitants, at least within the
confines of the solar system, to select their own celestial environments and change them as often as they
wished.

Whether such communities were stationary or
migratory, however, conditions in them—especially
if those communities were numerous—would favor
ethnogenesis, or the development, both through environmental conditioning and deliberate collective
choice, of new cultures and subcultures. And anthropologists, as community members, would be in an
ideal position to monitor such culture formation and
culture change while acting as participant-observers.

In sum, anthropology, as a young, far-ranging
and holistic discipline, is by its nature better pre-
adapted than most of the established arts and sciences to a central role in the impending exploration
of the extraterrestrial realm, whether extraterrestri-
ality is conventionally confined to interplanetary

space or anomalistically extended to include spatial, temporal and "metaphysical" domains of the type suggested by our species' rich but baffling mythology. In either case, it is to be hoped that most anthropologists do not succumb to the increasing academic pressure toward disciplinary overspecialization, leading them to neglect not only this cosmic perspective but even general anthropology in favor of kinship algebra or generative phonology. For their doing so would present us with the ironic spectacle of the science of man, once first in the intellectual battle against all forms of parochialism, succumbing to a newer and even more trivializing form of ethnocentricity.

Appendix

Anomalistics: The Outline of an Emerging Area of Investigation

Anomalistics may be defined as the serious and systematic (rather than sporadic and sensational) study of phenomena of all kinds which fail to fit the pictures of reality provided for us by common sense or by the established sciences.

For the anomalist, research of the sort conducted by academic, industrial and governmental institutions is, though necessary and important, always secondary to search—the initial effort which leads to the unearthing of fresh data or the framing of exploratory hypotheses. Such search is generated by skeptical imagination and guided by imaginative skepticism. Because it does not follow the pathways nor respect the boundaries of conventional disciplines, it is of necessity a generalized rather than a specialized enterprise.

From a short-range perspective, anomalistics seems to be in irreconcilable conflict with scientific orthodoxy. In

the long run, however, anomalism is of undeniable benefit to science, since such scientific "revolutions" as the
Copernican, Darwinian and Einsteinian have all resulted
from the exploration and exploitation of anomalous dis-
crepancies between observation and theory in earlier
systems.

The creators of anomalistics, in its explicit contemp-
orary sense, are the American journalist Charles Fort, who
made the first massive collection of inexplicable happenings from the world press, and the Scottish-American
biologist Ivan Sanderson, who founded The Society for
the Investigation of the Unexplained in Columbia, New
Jersey.

Anomalistics, like any other risky and controversial
endeavor, also has its martyrs. Two of the most recent are
Wilhelm Reich, the Austrian-American physician who
died in a federal prison after the burning of his books by
the Food and Drug Administration, and Immanuel Veli-
kovsky, the Russian-American psychoanalyst whose studies
of prehistoric global catastrophes made him a scholarly
outcast for twenty-five years.

Insofar as anomalistics may be regarded as a "discipline," it may also be said to contain "subdisciplines."
Among these are: parapsychology, the study of so-called
extrasensory perception and related phenomena; noetics,
the study of consciousness; and exo-biology, the study of
life forms presumed to exist on other planets. To these I
would add chronontology, the investigation of the nature
of time, as opposed to chronology, the systematic seriation
of events, and chronometrics, the study of time-keeping
methods and devices.

Perhaps the greatest single difference between anomalists and practitioners of established disciplines is that
the former question the uniformly progressive model of
human evolution assumed by most scholars. Anomalists
call attention to such embarrassing facts as the accurate
description of the moons of Mars by early eighteenth-
century writers—one hundred and fifty years before there
were telescopes powerful enough to descry them. They

also catalog such technological anachronisms as what appear to be depictions or remains of electric generators, computers, rocket launchers and jet planes from ancient Egypt, Greece, Rome and Colombia, respectively. And they remind us that in pre-Christian Egypt and Mexico, the earliest architecture and sculpture seem to be technically more sophisticated than what followed them.

In general terms, anomalists may be said to take seriously those views, reports and practices that most scientists dismiss as superstition. They decline to dismiss myths as mental aberrations or even to reduce them to distorted symbolizations of ordinary historical events or social relations. They call for a more careful consideration of astrology and alchemy—studies which absorbed most of the energy and attention of some of the most gifted scholars of the Old World for millennia. And while they rarely adhere to any of the religious orthodoxies, anomalists generally advocate a sustained and dispassionate effort to understand the so-called supernatural holistically, without attempting either to explain it away or to segregate it as inexplicably beyond the reach of investigation.

Among the more dramatic phenomena studied by anomalists are supposedly extinct or imaginary animals, such as ape-men or sea serpents, sightings of which are still periodically reported. Still more striking are the constant reports of mysterious lights in the sky, an ancient phenomenon which the early Romans referred to as "flying shields."

Perhaps the most outrageous of all anomalies are objects found embedded in rocks declared, on good geological authority, to be millions of years old: these range from iron chains to live frogs. Their only rivals as affronts to our conventional wisdom are objects fallen from cloudless skies, ranging from small jellyfish to gigantic ice blocks.

A more orderly way to arrange anomalies may be in terms of the academic disciplines under whose rubrics they fall. Examples follow:

1. astronomy: quasars (star-sized bodies radiating the energy of entire galaxies) .

2. meteorology: the Tunguska explosion of 1908 over Siberia (now shown to have been neither a comet nor a meteor and held by a prominent Soviet astrophysicist to have been a thermonuclear blast).

3. geology: tektites (glassy pellets found scattered in clusters in most parts of the land and sea).

4. oceanography: the so-called Bermuda triangle (actually one of ten lozenge-shaped areas in the tropics in which an unaccountably high proportion of ships and aircraft have disappeared without trace).

5. cartography: the Piri Re'is map (a pre-Columbian chart from Turkey delineating both Americas and the land—as opposed to ice—mass of Antarctica).

6. archeology: megalithic structures (from Stonehenge in England to Machu Picchu in Peru, whose construction, in the absence of wheeled vehicles and iron tools, would seem to have posed virtually insurmountable difficulties).

7. anthropology: the absence of Paleolithic Negroid fossils (suggesting that while whites evolved, blacks were created!).

8. physiology: fire-walking by non-Western adepts (whose feet, in our judgment, ought to get so badly burned as to cripple them for life).

9. ethnology: the occurrence in every mythology of tales describing world-wide fires and floods.

10. psychology: cryptesthesia (mental telepathy, clairvoyance, precognition and similar "paranormal" processes).

Notes

1. Personal communication with the author, based on Bateson's study of octopus behavior done at the V.A. Hospital in Palo Alto, California.

2. Burt and Ethel Aginsky, "The Domestication of the Species *Tursiops truncatus*" (Public lecture and film at Drew University, Fall 1968).

3. Carleton Coon, *The Origin of Races* (New York: Alfred A. Knopf, Inc., 1962), pp. 207–208.
4. Grover Krantz, "The Anatomy of the Sasquatch Foot," *Northwest Anthropological Research Notes* 6, no. 1 (1972) : 91–104.
5. John Napier, *Bigfoot: The Yeti and Sasquatch in Myth and Reality* (New York: Dutton, 1973) .

2.
First Contact with Nonhuman Cultures: Anthropology in the Space Age

DONALD K. STERN

Introduction

Until recently, man considered himself and his planet unique; he held both anthropocentric and geocentric views of the universe. But in fact we are far from being the center of the cosmos; we live on a small planet of an undistinguished star at the rim of the galaxy. We also now suspect that earth is not unique as an abode of life. Experiments have shown that complex molecules of the sort needed for primitive life to evolve can be made artificially under conditions similar to early planetary atmospheres.[1] This implies that life may regularly arise under suit-

able conditions on other planets. One of the major
problems with finding life in space is that we do not
really know what we are looking for, and may not
recognize it when we find it.

Since *Voyager* and *Mariner* it has become an
article of faith with many that "life as we know it" is
impossible anywhere in our solar system except on
earth. It is too hot or too cold, the pressure is wrong,
or the atmosphere is too poisonous to sustain life.
The assumption that since something as complex
and highly evolved as a mammal could not exist
under those conditions, "life as we know it" is impossible, is one of the most astounding examples of
ethnocentricity since the earth was considered the
center of the universe.

The problem appears to be more psychological
than scientific. What is usually meant by the hack-
neyed phrase is "life as I am *familiar* with it."
Those authors who enthuse about the exceptional
suitability of the terrestrial environment for life are
confusing cause and effect. The beginnings of life on
earth existed in an atmosphere of methane, ammonia, phosphine and hydrogen sulphide.[2] Even
today there exists one kind of bacillus that oxidizes
elemental sulphuric acid, using only a slight proportion of oxygen in the process. Another variety can
oxidize iron totally without the aid of oxygen. Some
organisms grow and reproduce only in oxygen-free
environments; in fact, oxygen is fatal to them.[3]

Sir Spencer Jones refers to the possibility, long
known to chemists, that silicon may replace carbon
in organic molecules and so enable life to exist at
temperatures above those usually considered favorable.[4] J. B. S. Haldane makes passing mention of
liquid ammonia replacing water as an organic sol-
vent in frigid planetary environments.[5] And Shapley

says, "There is the fanciful possibility that a planet with life could be independent of the gravitation and radiation of a star . . . with its necessary heat generated by its own radioactive mineral constituents. Such a planet would have no proper day or night; it would be entirely dark except for the weak starlight and luminescence incited by the radioactivity; it would have a strange atmosphere, be devoid of tides, and have a most unusual climate and certainly a most exotic biology."[6] As we know from life in our oceans, total darkness is no bar to life, nor is a solid surface an absolute necessity. Life on earth most probably originated in the sea, so a wholly aquatic planet could be a possible "abode of life," to use Percival Lowell's phrase. And a very dense atmosphere does not differ essentially from an ocean.

Radio Contact

Probably radio contact is safest as far as man's ethnocentric mental well-being is concerned. This is also the most likely form of contact to occur based on what we consider to be the limitations of space flight, and is the easiest to direct and coordinate.

If we assume that life does exist elsewhere in the universe, then it would seem reasonable that some forms of life have developed a technology at least as advanced as ours, if not more so. After all, we have only had technology a short while; other worlds may have had it for eons. Intelligent beings might not always be technological, nor might they wish to communicate with other worlds. But if there are other civilizations attempting to communicate across space, we can probably assume that they have a system of mathematics, some means of observing the heavens, and a language capable of abstract concepts.

The search for signals from extraterrestrial civili-

zations is based on radio frequencies which provide
the optimum range for transmission of meaningful
signals over large distances. At a certain level of
technology, civilizations inevitably start transmitting
information into space. It is assumed that such a
broadcast would be related to the experimental exploratory activity of an advanced technological
civilization trying to locate its fellows.[7] This, of
course, supposes that the psychological/ethical
trends of a highly organized society generate a certain pressure for the transmission of signals into
space,[8] and that we can decode these signals.

We cannot guess what the character of an advanced extraterrestrial civilization would be. But the
laws of physics appear universal in character, and if
familiarity with electromagnetic theory is common to
all technological races in the galaxy, is it possible to
establish radio contact between them? If contact
were established with a race possessing a high degree
of scientific development, the impact on our lives,
society and philosophy would be incalculable.

It is generally agreed that the idea of communication with extraterrestrial civilizations passed from
the realm of science fiction to that of science in 1959,
when Cocconi and Morrison suggested that the signals of these civilizations should be sought at the
natural wavelength standard, the 21 cm. radio line
of atomic hydrogen.[9] But this is not the only possibility. In our search, we are faced with a twofold
uncertainty: we do not know at what frequency or in
what direction these signals are to be sought. A similar uncertainty exists for the other civilizations.
Before establishing communication, the sources of
these signals must be reliably identified and distin-
guished from a tremendous number of natural radio
sources.

The simplest solution to this difficulty is the transmission of signals of a sufficiently wide band. The greater the band width of the communication line, the higher the given number of signals that can be transmitted simultaneously. Each signal is associated with a certain message, characterized by a definite quantity of information.[10]

The difficulties of electromagnetic communication over interstellar distances are immense. Calculations suggest that the limit of reception of beamed transmissions and dish aerials is about a hundred light-years. If the transmissions were not beamed, the range would be much less. The prospects for interstellar communication over distances of tens of light-years seem reasonable; over larger distances, they range from difficult to nearly impossible for someone at our present level of technology. If it seemed likely that technical societies existed only ten to twenty light-years away, a serious effort to establish contact might be justified. On the other hand, if we can only expect to find them at distances of thousands of light-years, attempts to communicate at this stage are futile and impractical.[11]

One of the most probable elements of activity of these "supercivilizations"[12] would be the transmission and exchange of information. These transmissions can be divided into two broad types: (1) exchange of information between highly developed civilizations of approximately the same level, and (2) transmission of information aimed at raising the level of less developed civilizations. If supercivilizations actually exist, transmissions of the first group would be inaccessible to us, since they would probably be by tight-beam. On the other hand, those of the second group would be readily accessible and easily detectable.

Initially, the transmission of words in the language of the transmitting civilization, no matter how simple, would not be useful for interstellar communication. Each race has a different evolutionary history, and may exist on a planet with an entirely different physical environment. Its thought patterns, social conventions and cultural values may be entirely alien to our own. Thus we would be receiving a word without a cultural context and it would be difficult to discern its precise meaning.

Von Hoerner has suggested that signals from another civilization would ultimately be determined by the purpose they are to serve, and by the most economical transmission channel.[13] D. C. Holmes points out that there are three general types of signals which might be employed: local radio traffic, beamed transmissions to regular "correspondents," and beacon signals intended to attract the attention of any civilization not yet contacted.[14] The first, although faint, might be picked up by another civilization, interpreted as the product of a technological civilization, and beamed back to its origin as a recognition signal followed by a message formulated in "interstellar linguistics." Some years in the future we may be receiving our "Inner Sanctum" radio show from Capella.

A beamed transmission intended for someone else would be detected by us only if our planet were accidentally interposed in the direct radio path between two communicating galactic civilizations. The probability of this interception is very slight.

We would only expect, then, to hear the beacon signals. Communicating beings would probably be familiar with mathematics and certain fundamental physical constants, so that we might expect a prime number sequence, pi, or the ratio of the mass of an electron.

The aim of any communication system is the transmission of certain messages. These may constitute written text using the letters of a particular alphabet (as in Morse code) or sounded verbally (as in radio). The messages may also constitute an image (as in television) or an algorithm to be transmitted to an automatic control system. Any of these messages may be represented as a series of digits.

We can advance a number of assumptions regarding the likely composition of extraterrestrial call signals. First, to ensure a high detection reliability, they would use continuous radio transmission. In addition, they would contain at least a minimum quantity of information labeling them as artificial sources, indicating frequency and band width of transmission, and some additional information which may be regarded as "key" to the main program.

A basic problem in decoding messages is the virtually total lack of any prior information or knowledge about the transmitting civilizations. Essentially, we are faced with the problem of decoding an arbitrary text.

The transmission of a core of knowledge about a civilization would most likely be preceded by a linguistic introduction signal, which in turn would be preceded by an announcement/recognition signal. Preliminary information could be transmitted graphically. But a nongraphic "interstellar vocabulary" has been developed by the Dutch mathematician Hans Freudenthal for the expression of more abstract ideas. This artificial language, called LINCOS, is designed as an entirely logical nonverbal

language free of the inconsistencies of grammatical
rules and other irregularities found in spoken
language.[15]

While the study of terrestrial languages includes
grammar, syntax and phonemics, LINCOS is designed entirely in terms of semantics. It consists of a
coded system of units which are clearly enumerated
into chapters and paragraphs. This facilitates the
interpretation of the message, because semantic content can be derived from logic external to the linguistic system itself.

A transmission in LINCOS begins with the most
elementary concepts of mathematics and logic. This
is because the language must define itself before it
can become a system of communication.

During a course in mathematics, a recipient
would be introduced to such concepts as "more
than/less than," "similar to/different from," and
"maximum/minimum." Each of these concepts
would be useful in deciphering subsequent information. According to Freudenthal, LINCOS could also
transmit more complex ideas which characterize human nature, such as "cowardice," "anger," or "al-
truism" by transmitting short theatrical perfor-
mances conveying emotions, social conventions and a
wide range of philosophical considerations.

Linguistic information transmitted alternately
with pictorial information could be especially effective, particularly in the transmission of scientific
data. Mendeleev's periodic system of the elements
could be pictured, accompanied by the correspond-
ing words in LINCOS.

While it seems likely that the transmissions of
pictorial representations and artificial languages
such as LINCOS would be easily understood by an
alien mentality, this is really conjecture. We do not

know what hidden assumptions lie in our proposed communication channel, assumptions which we are unable to evaluate because they are so intimately interwoven into our fabric of thought. It remains to be seen whether mathematics is an interstellar Rosetta stone.

In comprehending an extraterrestrial's message, there should be little difficulty in understanding the math and physics. But there is no guarantee that the concepts of enmity and friendship, learning and modes of information exchange, and observation and experimentation are universal and not merely anthropocentric. The problems start with terms like *I* and *mine*. Ever wonder what a bee's concept of *I* might be?

Our present scientific knowledge enables us to analyze the conditions of signal transmission through interstellar space, to consider the requirements to be met by the transmitting, and especially to consider the receiving systems and antennas. The main problem falls into two different categories: the direct search for signals and reception of information from extraterrestrial civilizations.

But electromagnetic communication does not permit one of the most interesting categories of interstellar contact—namely, contact between an advanced civilization and an intelligent but pretechnical society. Such contact would be potentially valuable, because the lifetime of the pretechnical era on many planets may be quite long, and the number of these societies may exceed by far the number of those technically advanced.

Neither direct exploration of the interstellar medium with its wide range of physical phenomena unobservable from the solar neighborhood, nor the examination of extraterrestrial biologies, nor the

direct exchange of material objects among distant
civilizations is available with only electromagnetic
communication.

Automatic Probes

In general, the development of a technologically
advanced civilization will be accompanied by great
progress in space vehicle technology. Relatively early
in its technical lifetime, the civilization would be
capable of sending small automatic interstellar
probes containing exceptionally long-lived radio receiving and transmitting apparatus to the nearest
stars. The energy required to power this apparatus
could come from the star being orbited.[16]

There are a number of distinct advantages to this
type of contact. Once in orbit around the local star,
the probe would automatically attempt contact with
planets in the vicinity. Since the instrumentation
would be powered by the star, its signal would be
more powerful than one sent from earth, and would
have a shorter distance to travel.

Physical contact effected by relatively short-range
interstellar probe vehicles would have some interesting properties. It is conceivable that physical objects
could be transported in such vehicles to civilizations
circling neighboring stars.[17] The exchange of cultural artifacts would have a beneficial influence on
maintaining contact. If interstellar voyages over distances greater than tens of light-years are undertaken, this artifact diffusion will provide a
connecting link between civilizations. An artifact of
such incredible beauty or power might evolve into
an object of worship even in a technical civilization.
Similar circumstances and an entire mythology have
developed under analogous conditions in the con-
temporary cargo cults of New Guinea—an example

of artifact diffusion contact between civilizations of greatly differing levels of technology.

In 1972 the United States made an attempt (and hopefully not an abortive one) at sending an un-manned probe on an interstellar journey. After pass-ing within 87,000 miles of Jupiter, the *Pioneer 10* probe radioed back the first close-up pictures of the giant planet, and probed its intense magnetic fields and radiation belts. Then, with the aid of the planet's powerful gravitational field to act as a sling-shot, the vehicle was hurled beyond Jupiter to begin the first voyage of a man-made spacecraft outside of the solar system.

Cornell astronomers Carl Sagan and Frank Drake persuaded NASA to attach a plaque to the probe's antenna supports to indicate where it came from and who its builders were. The six-by-nine-inch alumi-num plaque was anodized with erosion-resistant gold, and the symbols etched into it were designed to be meaningful even to beings totally unfamiliar with human logic.

As their central illustration, Sagan and Drake chose figures of a man and woman. Their height is indicated by the scale drawing of *Pioneer 10* in the background. The message also contains an illustra-tion of a hydrogen atom, the most abundant element in the universe, undergoing a change of energy state. During this process, the atom gives off a pulse of radiation with a 21 cm. wavelength (which is also the message's basic unit of measure). The message's most interesting feature, however, is a large starburst pattern. Fourteen lines symbolize specific pulsars, each supposedly recognizable by the precise fre-quency, noted in binary terms, at which they give off radio signals. A fifteenth line extending behind the humans indicates the distance of our sun to the

center of the galaxy. That information should tell
extraterrestrial scientists when and where the vehicle
was launched. For specific details, there is a repre-
sentation of the solar system at the bottom of the
plaque showing the route taken by *Pioneer*.

In planetary probes and more sophisticated
future interstellar probes, the list of proposed life-
detection systems is quite large and impressive.
Sneath includes the following: (1) televising the
planet's surface for visual observation of characteristic forms of living things; (2) microscopy of surface
constituents for detection of microorganisms; (3) gas
chromatography and mass spectrometry of heated
soil samples for identification of complex organic
molecules; (4) J-band detection of changes in the
absorption spectra of certain dyes when they react
with large molecules of biological origin, like protein
or DNA; (5) liberation of radioisotopic CO_2 from
culture media (microorganisms should attack simple
common compounds and liberate radioactive CO_2);
(6) measuring microorganism growth in the media;
(7) testing for biochemical reactions for specific sub-
stances associated with living organisms.[18]

Most life-detection experiments assume that any
indigenous life will behave like terrestrial organisms.
To look only for well-known life systems guarantees
that we can evaluate any results with confidence. But
it would be unwise to rely on this alone, or to assume
that negative results rule out any kind of life.

Direct Contact

"There is a high probability that civilization is a
universal phenomenon, and yet there are no currently observed signs of cosmic activity of intelligent
creatures."[19] Indeed, the available data on the conditions for the evolution of life and the number of

probable planetary systems suggest that life is a fairly commonplace and regular occurrence in the universe. There are several billion planetary systems in the galaxy, and of them about a billion worlds may be populated with their own varieties of living organisms.

The ascent of life has followed two courses: specialization, allowing the best to be made of a particular environment; and "universalization," consisting of flexible adjustment (reversible adaptation) involving choice and nonrandom action. This requires perception, coordination and cognition. In the next stage, the environment itself is adapted to the needs of the organism by its own efforts, and finally by collective action in an organized community.

On some planets, life may have existed for such a long time that there may have arisen a series of technologically advanced civilizations. There is only about a 0.5 percent probability that any given interstellar contact would be with a civilization at the same level of development as our own. Thus in any contact at present, we very likely will have more to learn than to teach.

In interstellar radio communication, the participants are distant, the learning vicarious, and the duration of the discourse overlong. But if interstellar flight were possible, it would remove these difficulties. The possibility of automated interstellar flight has already been discussed. It is now necessary to examine the prospect of direct contact.

If interstellar flight is technologically feasible, albeit expensive and difficult, it is likely to be developed by a civilization substantially in advance of our own. Even beyond the exchanges of information and ideas with other intelligent races, the scientific advantages

of this type of flight are immeasurable: direct astronomical samplings of distant planetary systems and
stars in all stages of evolution, and the observation
and sampling of a multitude of extraterrestrial cultures and biologies.

Direct contact and exchanges of information and
artifacts should exist among most spacefaring societies possessing starships. The situation bears some
resemblance to post-Renaissance seafaring communities and their colonies. If fast, inexpensive interstellar transportation is feasible, the technical civilizations of the galaxy will be an intercommunicating
whole.

The Soviet astrophysicist N. S. Kardashev, in
speaking about extraterrestrial societies, devised the
categories of Types I, II and III as measured by the
capability to control greatly different orders of mag-
nitude of energy.[20] Type I civilizations control the
resources of their planet, and resemble our own. The
only method for detecting them is through radio
emission. Type II civilizations use a large percentage
of the energy of their parent star. These societies
were first examined by the American physicist Free-
man Dyson. By making a vast swarm of "solar cell
particles" from the matter of a Jovian-sized planet,
and redistributing them to form a thin spherical
shell around both star and planet, this "Dyson
sphere" would make such a civilization appear as an
intense infrared object. By contrast, a Type III
civilization should be conspicuous, since it would
theoretically control the resources of an entire
galaxy. Looking for one has been likened by Dyson
to "searching for evidence of technology on Manhat-
tan Island."[21] But would a primitive recognize a
technology if he saw one? In a science fiction story by
Howard Waldrop, an automated probe was mistaken

for an unusual life form. When it did not respond to a native's overtures, it was "killed."[22]

The difference between energy consumption levels of Types I and II and Types II and III is a factor of about 10^9. This is a large number, but "a society growing at the rather modest rate of 1% per year will go from Type I to Type II in less than 2,500 years. Going from Type II to Type III will take longer, since it requires that the race spread out among all the stars. Possibly this could be accomplished within 100 thousand years . . . all these are just linear extensions of our own civilization."[23] In a way, however, the whole idea of detecting a superior technology by its waste heat is absurd. It is a little like trying to analyze another culture by its garbage.

The development of a civilization is usually treated in terms of sociopolitical and economic development, evolution of language and art, development of science and technology, and the role of religion. We cannot maintain, however, that these factors as we understand them now will apply indefinitely to describe the process of Terran civilization. Moreover, there is no a priori justification for extending these concepts to other civilizations, constraining them to follow approximately the same evolutionary course as ours.

The development of a civilization is largely determined by the abundance of natural resources. Planets of stars where fossil fuels are plentiful and where heavy metals such as iron, copper, tin, silver and gold are abundant and available could be sites for highly advanced civilizations. It might be possible to build a complex technology out of nonmetals or light metals, but it does not seem likely. Tribes on earth that never had easy access to heavy metals never developed a high technology. And light metals are much less abundant in the universe than the heavier metals because they are excellent fuel for the nuclear fusion reactions that power stars. It is the heavy metals that form a natural link with the forces of magnetism, electricity, gravity and nuclear energy.

But intelligence per se does not depend on these resources. It has been emphasized that the development of what is referred to somewhat loosely as "mind" is an aspect of the general process of evolution.[24] Its development is part of general organic development. Survival depends on the ability to perceive, evaluate and actively influence the relationships between the organism and its surroundings. It confers on a species the power to live in a greater range of habitats, and consequently to survive greater dangers of natural catastrophe. In fact, according to H. S. Jennings, even simple organisms exhibit a type of awareness and intelligent adaptation. He described the behavior of an amoeba in terms of "pursuit, capture and ingestion of one amoeba by another, the escape of the captured, its recapture and final escape."[25]

The dictionary definitions of intelligence are, variously, "the faculty of reason," "the capacity to adapt to new situations," "the power of understanding," "the ability to acquire and use knowledge," and "the ability to invent, create, and imagine." By these criteria, many animals besides man show a degree of intelligence. It might be better, therefore, to think of intelligence in terms of different aptitudes involving planning (forethought), memory (afterthought), learning (adaptability) and creativity (imagination, image- or pattern-forming). There should be some minimum evidence that a creature has all of these abilities before terming it intelligent; or more precisely, that the level of intelligence is primarily determined by the ability that is least well developed.

Samuil Kaplan, a noted Russian astrophysicist, states intelligence as "a highly stable state of matter capable of acquisition, and abstract analysis of the maximum quantity of information about the environment and itself, and developing survival reactions."[26] And while intelligence may be an actual and identifiable entity, our sample of "intelligent life forms" is a very narrow base for conclusions. We might not recognize what another race considered "intelligence." And a third alien would probably do no better.

Intelligence requires an elaborate brain. This is not entirely a matter of size. The brain of an elephant is larger than that of a man but shows specialization of a low type, with fewer convolutions. Whales, however, present a striking contrast: their brains are both large and convoluted. Both whales and dolphins are exceedingly intelligent and are the subjects of a great deal of current research.

The American neurophysiologist John C. Lilly of the Communications Research Institute has agreed that dolphins and other *Cetacea* have surprisingly high levels of intelligence.[27] Dolphins are capable of making a large number of sounds of great complexity, which almost certainly are used for communication. Recent evidence suggests that they are capable of counting, and can even mimic human speech. They have very limited manipulative abilities, and despite their level of intelligence could not have developed a technical civilization. But their intelligence and communicativeness strongly suggest that these traits are not limited to humans.

It would be hard to argue convincingly that all higher intelligences in the universe must be technological. Rather more plausible is the argument that intelligent beings must be social beings, because "only a society will produce the necessary teachings and the conflicts of interest that goad minds into thought."[28]

Insect societies undoubtedly constitute a development both akin to and different from human organizations. The activities of social insects are not automatic.

> So often a little hitch occurs, something in the usual chain of events fails, then you may see the insects changing their tactics and making trials until something succeeds . . . The astonishing thing about ants in such dilemmas is that they usually hit upon the only solution to their problem in such a short time. This is generally the work of an individual leader, and the community will follow and complete the task.[29]

The social insects do not use tools or machines; they grow them. They produce individuals with strong jaws or with special glands that secrete corrosive chemicals or with overdeveloped ovaries. This is not a matter of heredity, but the way in which the grubs are reared, rather like the Alphas, Betas and Deltas in Aldous Huxley's *Brave New World,* only more so. "A creature well-fitted to its environment needs no tools. It is only the weak, the unfit that change the world."[30]

Within a body as small as an ant's there is not enough nervous tissue for any sort of highly developed brain. The problem has been resolved by the formation and evolution of a collective mind along the lines of William M. MacDougall's "group mind" in human associations.[31] The difference is only that in the latter, the collective mind is largely externalized, being contained in books, records, means of communication and diffusion of information, instruction, entertainment and propaganda.[32] In illiterate societies, however, culture and tradition are passed orally from generation to generation. The individuals involved have memories of prodigious retentiveness to cope with this task, a faculty that decays with the spread of literacy.

If intelligence always evolved, given enough time, we could expect a high incidence of intelligent life on other worlds. On the other hand, it may require a unique combination of circumstances to encourage this development. After all, the assumption that technical civilizations must necessarily appear after many billions of years of biological evolution implies that the ultimate purpose or goal in the formation of stars and planets is the production of intelligent beings and technical civilizations. This viewpoint is both idealistic and teleological. Life had been evolving for 3,000 million years before the human race appeared; birds and mammals took only 200 million years.[33]

Finding life beyond the earth, particularly intelligent life, wrenches at our secret vision of man as the pinnacle of creation, a contention which no other species on earth can challenge. Even simple forms of extraterrestrial life may have abilities and adaptations denied us. Perhaps the greatest blow to man's ego would be to meet an intelligent extraterrestrial race that also considered itself to be human. The discovery of life on some other world will, among other things, be for us a humbling experience, since man's image of himself seems to require an anthropocentric god.

The power of speech and mental abstraction are usually held to be a distinctive achievement of mankind, setting it apart from the rest of the animal kingdom. While it is true that other vertebrates have not progressed beyond phatic expression (communicating emotional states), much of human speech in everyday life does not go far beyond this stage, either. Human speech, in most cases, serves to convey little more than affects and attitudes.[34] A certain amount of abstraction enters into phatic communication, and it has even been shown that other organisms can count. The Venus flytrap, for instance, has trigger-hairs on its traplike leaves which close on its prey when touched. The first touch will not operate the trap, but a second touch will. It can therefore count accurately up to two.[35]

"Extraterrestrial contact" necessarily precludes the fact that a civilization knows it is being contacted. If "undercover agents" or "guardians" are planted within a society without the knowledge of the natives (presupposing a similarity in genetic and evolutionary stock), contact as such is not realized. Therefore, certain restrictions must be placed on what constitutes a contact.

The establishment of contact with extraterrestrial civilizations may not only lead to radical changes in our basic concepts regarding an intelligent society, however "logical" these concepts appeared to be prior to contact, but also may greatly affect the future development of our own civilization. This is a result of the feedback effect often discussed in connection with the beneficial or harmful outcomes of interplanetary/interstellar encounters. It is difficult to foresee how the rates of our progress would be affected by direct contact with the representatives, living or mechanical, of other civilizations. Science fiction writers advance a variety of hypotheses regarding the impact of this encounter on the scientific and social advancements of the more backward of the two societies. The result of this meeting is the feedback effect often discussed in connection with the beneficial or harmful outcomes of interplanetary/interstellar encounters. Some feel that no significant change in the rate of progress can be brought about by this intervention unless the recipient society were to lose its individuality.

There are no reliable reports of direct contact with an extraterrestrial civilization. Any early accounts are encumbered with some degree of fanciful embellishment, due largely to the prevailing views of the time. There are some who think that around 15,000 to 20,000 B.C. the earth was visited by a race of interstellar "supermen" who left behind them all the myths of the Olympian gods and goddesses. "Who were the Titans, deformed giants like the one-eyed Cyclops, or the hundred-handed Briareus? Were they slaves of a lower class, mutants born of parents and grandparents forced to live and work in proximity to the hard radiation shed by a starship's engines? We shall never know."[36] In a similar vein is the inexplicable story in the first chapter of the Book of Ezekial, in which the prophet describes how he saw a quartet of "four-faced men" emerge from a whirlwind of "enfolding fire." With "the likeness of men," these four had "four wings" over their heads that turned into wheels and lifted them; ". . . and their clothes flashed in the sun like burnished brass."

To the medieval translators of the Book of Ezekial, such impossible creatures meant nothing. To the modern world, Ezekial seems clearly to have described with remarkable exactness four beings in woven-metal spacesuits, with some sort of portable, lightweight helicopter units strapped to their backs. When the blades spun, they turned into wheels to "the noise of great waters," while "their rings were

full of eyes"—not unlike the shock-wave pattern set
up by supersonic tip-jets. As for the whirlwind of
enfolding fire, the accompanying cloud and the
"brightness about as of amber," a descending rocket,
standing on its fiery tail, might look much the same
from a distance.

Another more relevant incidence is the native
account of the first contact between the Tlingit
people of the northwest coast of North America and
European civilization—an expedition led by the
French navigator La Perouse in 1786.[37] The Tlingit
kept no written records, but one century after the
encounter, the verbal narrative was related to the
American anthropologist G. T. Emmons by a principle Tlingit chief. The oral rendition contained
sufficient information for later reconstruction of the
true nature of the contact, although many of the
incidents were disguised in a framework of mythology—for example, the French ships were described as
large black birds with white wings.

This encounter suggests that under certain circumstances, a brief contact with an alien civilization
will be recorded in a reconstructible manner. This
reconstruction would be greatly aided if (1) the
account is committed to writing shortly after the
event; (2) a major change is effected in the contacted society by the encounter; and (3) no attempt
is made by the contacting civilization to conceal its
exogamous nature.[38] A description of the morphology of a nonhuman, or a clear account of astronomical data which could not be acquired by a primitive
people, would increase the credibility of a legend.

Unfortunately, even present-day sightings of unidentified flying objects and their occupants are
clouded. "Little green men" may be neither little,
green, nor men. The fact that most alleged sightings

of extraterrestrials turn out to be decidedly human, and on occasion supposedly from another planet *within* our solar system, further accentuates either our ethnocentrism or our gullibility. This is not meant as a dismissal of *all* sightings, humanoid or otherwise, but as Carl Sagan notes, there have been no reports of flying saucers from astronomers, who watch the sky almost constantly and photograph it extensively.

What might an advanced extraterrestrial civilization want of us? It has been postulated that a space-faring society capable of crossing interstellar distances would be comprised of wise and benign beings. The same might have been said of the Europeans during the Renaissance period of exploration, when ocean voyages were on par with today's exploration of space. "One of the primary motivations for the exploration of the New World was to convert the inhabitants to Christianity—peacefully, if possible; by force, if necessary. Can we exclude the possibility of interstellar evangelism?"[39] The American Indian was not useful for any concrete task in the courts of France and Spain; they were transported there as objects of curiosity and for prestige purposes. Would this be an emotion unknown to aliens?

If an alien vessel entered our solar system, but elected to take up a parking orbit some distance from the earth (say about midway between Mars and Jupiter), our first step, if radio communication failed, might be to send a probe in an attempt to gain more information about the intruder. As a result, there would be relayed back to earth pictures of a vehicle orbiting like some tremendous battleship cruising off the coast of some tiny, backward island. Inevitably, there would be those who wanted a closer look. This being the case, one or more "dugout canoes" would probably be readied to send a small delegation of "natives" to greet the stranger.[40]

A project group such as Ozma might be put in charge of both radio and physical contact, necessitating the development of new departments to meet the increased flow of information acquisition, utilization and diffusion. Such departments might be known as Communications and Cryptography; Culture and Customs; and various science divisions concerned with both Terran and extraterrestrial biology, chemistry and physics.[41]

If the cultures are in close enough contact, one race might record a scientist of the other conducting an experiment, the purpose of which would be to learn the referents of their symbolic structure that describe their concept of scientific method. For the "aliens," science might not be a description of reality so much as a metaphorical ordering of experience.

Knowledge of the possibility of impending contact with what are known to be nonhuman life forms (through radio communication prior to physical contact) could create widespread xenophobia. The psychologist C. G. Jung observed that contact with superior beings might be shattering to us; to find ourselves no more a match for them intellectually than our pets are for us, to find all our aspirations outmoded might leave us completely demoralized.[42] This could be typified in many ways. Plots by various extremist groups might be established to kill the aliens.

Perhaps a coordinating project like Ozma would push too many new ideas and technologies too fast. Culture shock would occur as rapid technological changes were introduced through the interpretation and release of extraterrestrial science; there would be too strong a challenge to basic Terran traditions,

customs and mores, brought about by the impinge-
ment of a new species' necessarily different viewpoint. If the project were actually the source of a
society's break with its own culture, tradition and
mores, then perhaps it should be cut back or slowed
down. But what of the potential loss involved?

By the time we are capable of interstellar travel,
we might stumble onto another spacefaring race "out
there." The first reaction to this accidental contact
might be that both parties would turn and bolt for
home. Realizing, in their own xenophobic way, that
the other ship might follow them to their home
world, they would be forced to establish relations of
one sort or another. Should they open negotiations
with the other vessel, or are they duty-bound to
attack and destroy it without warning? In Murray
Leinster's "First Contact," the humans and aliens
were enough alike biologically to exchange ships and
head for home by circuitous routes.[43]

Many take it for granted that the meeting of two
civilizations that have sprung up in different parts of
the universe is bound to lead to the subordination of
one by the other, that a meeting in space could only
mean two things—trade or war. The history of human civilization is the history of conflicts between
nations, societies and races. Could human beings
build a viable society with nonhuman partners?
What would happen to those youngsters who wanted
to adopt alien ways and desired to be something they
physically cannot? Questions such as these open the
door to a unique aspect of culture shock.

But what of extraterrestrials based on other
chemical and biological systems? If they are totally
different (chlorine-breathers, silicon-based instead of
carbon-based, ad infinitum), any relations would
definitely start off handicapped. But once the lan-

guage barrier was removed, relations could turn out
to be mutually beneficial and complementary.
Different environments would mean different abilities and skills; different planets would contain
different mineral wealth. Trade could be set up on
the basis of which minerals each culture needed and
could utilize from the other. This contact might also
involve various aspects of exploration and research,
such as each species working for the other in planetary environments suitable only for one.[44]

An anthropocentric attitude is a great handicap
in the study of sensory perception in differently constructed organisms, and is even more restrictive when
it comes to understanding their mental processes. It
is useful to remember that ordinarily we cannot be
directly aware of the mental processes of members of
our own varied cultures. Outside of the human kind,
even the slender aid of psychological insight forsakes
us, and the further away we move from man's
physiological make-up, the less we have to go by.

It is clear that the senses must be based on natural phenomena commonly occurring in the organism's surroundings significant to the organism's needs:
communication for such purposes as sexual reproduction, care of the young, pursuit of prey, escape from
danger, and with progressing collectivization, common defense and attack, and the search for or production of food by the pack, flock or society, terrestrial
or otherwise. Communication may be by sound, touch,
scent, light gesture, electrical pulses, and perhaps
other means for which we have no name.

The great variety of existing language families on
our planet and the absence of any obvious relationship between them can be attributed to the features
of Terran topology: mountains, distance by sea,
separate land masses and so forth. On a planet with

a more compact dry-land area, fewer mountains and deserts, and a greater number of navigable rivers, a single universal language would be more likely to develop. This development would be aided on both worlds by the availability of efficient systems of transportation and communication.

However, if a race happens to consist of visual telepaths, there would be no spoken language as such. Instead, if vocalizations were used, they would probably be only for identification purposes, such as recognition signals.[45]

Exobionts will no doubt possess various bodily organs just as we do. Their bodies will have to perform similar functions to our own, and these functions will be most efficiently carried out by specialized tissues. They can be expected to have digestive and excretory systems, a transport system to distribute nutrients throughout the body, and specialized organs to facilitate movement, biosynthesis and reproduction. Although it is probable that not all inhabited worlds have developed along lines resembling ours, if any intelligent life is to evolve, one would expect a similar set of evolutionary pressures to operate. Their world would probably have such familiar ecological classes as carnivores, herbivores, parasites and symbionts.

There are many ways of perceiving one's surroundings. Human life is strongly oriented towards the visual. Other animals have other dominant senses. Bats rely almost entirely on sonar; they hear the reflection of sound waves emitted at a high frequency pitch. Even more extraordinary are certain fish related to the electric eel. Instead of producing an electric shock to stun its prey, the Gymnarchus uses weak electric discharges to sense the whereabouts of other fish. There is also evidence that some animals are sensitive to magnetism or radio waves, and there seems to be no theoretical reason why this cannot be so. Rattlesnakes have infrared detectors that allow them to locate their prey. Perhaps they have a heat picture as well as one of light.[46]

In any human community, earth-based or otherwise, certain contact procedures must be followed. Since culture constrains perceptions of new situations, these situations and objects tend to be described by analogue. We must therefore be on our guard against the tendency to ascribe human thoughts and emotions to nonhumans. While it is usual in animal psychology to observe the behavior of an organism in response to simple tests, such artificial simplicity tends to falsify the issue. If a human were subjected to such tests as pain response, electric shock and unpleasant chemicals by some nonhuman investigator, the elicited reactions could easily be accommodated within the category of reflex and taxis.

Another incompatibility aspect might be controlling one's reactions to an alien's appearance. Being upset by certain visual stimuli is not a matter of reason so much as a physiological reaction. It can be controlled, but the strain is there, and aliens would probably feel the same about us.

Being overly concerned about another's taboos can severely limit subject matter open to discussion; all the things about them that strike us as peculiar or distasteful, and everything they might consider taboo. This "cultural timidity" would make us agreeable to opening any topic so long as *they* started it first. But what if they had the same idea? There would be no chance for a language barrier or communication breakdown, because true communication would never get started. An understanding must be reached: their habits are right for them, and ours for us, but we would not necessarily care to trade them.

If contact became both long-term and relatively frequent, there is a high probability that exobiotic medicine would become a standardized profession. To facilitate classification of life forms from a medical standpoint, a system such as that used by James White in his "Sector General" stories might well be used.

> Unless you are attached to a multienvironmental hospital, you normally meet e-ts one species at a time and refer to them by their planet of origin. But here, where rapid and accurate knowledge of incoming patients is vital, because all too often they are in no condition to furnish this information themselves, we have evolved the four-letter classification system. Very briefly, it works like this.
>
> The first letter denotes the level of physical evolution . . . The second indicates the type and distribution of limbs and sense organs and the other two the combination of metabolism and gravity-pressure requirements, which in turn gives an indication of the physical mass and form of tegument possessed by a being. Usually we have to remind some of our e-t students at this point that the initial letter of their classification should not be allowed to give them feelings of inferiority, and that the level of physical evolution has no relation to the level of intelligence.[47]

Species with the prefix A, B, or C were water-breathers. On most worlds life had begun in the seas, and these beings had developed high intelligence without having to leave it. D through F were warm-blooded oxygen-breathers, "into which group fell most of the intelligent races of the galaxy," and the

G to K types were also oxygen-breathing but insec-
tile. The Ls and Ms were light-gravity, winged
beings.

Chlorine-breathing life forms were contained in
the O and P groups, and after that came the more
exotic, the more highly evolved physically, and the
downright weird types: radiation-eaters, frigid-
blooded or crystalline beings, and entities capable of
modifying their physical structure at will. Those
possessing extrasensory powers sufficiently well-
developed to make walking or manipulatory ap-
pendages unnecessary were given the prefix V
regardless of shape or size.

> There are anomalies in the system, . . . but those
> can be blamed on a lack of imagination by its origina-
> tors—the AACP life-form, for instance, which has a
> vegetable metabolism. Normally the prefix A denotes a
> water breather, there being nothing lower in the
> system than the piscatorial life-forms, but the AACPs
> are intelligent vegetables and plants came before fish.[48]

Habitable planets lacking technical civilizations
will frequently be encountered by spacefaring societies. The latter may wish to leave such cultures
strictly alone and allow them to slowly evolve in
their own fashion. Direct contact may be delayed
until the natives develop a technical society at their
own pace. Strict injunctions against contact and
colonization of populated but pretechnical planets
might be put into effect. Or perhaps attempts would
be made to colonize every habitable planet without
regard for the indigenous inhabitants for purposes of
prestige, exploitation or some nonhuman motivation
which we cannot even guess at. A whole spectrum of
intermediate cases can also be imagined, in which
small colonies are planted on these planets, not to

interfere with or direct the development of the local life forms, but merely to observe them. If colonization is the rule, then even one spacefaring civilization could spread throughout the galaxy at a relatively rapid rate, on the galactic time scale (see Type III civilizations). There would be colonies of colonies of colonies, such as arose at many sites in the western Mediterranean during classical times.

One of the basic tenets throughout science fiction is the avoidance of cultural interference and contamination of less advanced cultures (usually in the technological sense) by "superior" ones. In many cases, merely discovering that an advanced civilization exists would be enough to alter the fabric of a culture by destroying the necessary smooth and logical progression in the discovery and development of ideas. And if a culture had an unfortunate experience in dealing with another race, it might impose drastic and radical precautions to guard against trouble with races not yet imagined. An alien technology may be one not easily recognizable, no matter how simple or sophisticated it may be. "Obvious savages" may have a technology built on a biological rather than a mechanistic basis. Their entire culture would depend on their ecological system running smoothly. The government would probably be continent-wide, since ecological interactions interlock strongly over an entire land mass. The coming of off-worlders could upset the system enough to collapse it. Thus the major difficulty is in correctly assessing a nonhuman culture with its equally nonhuman values, especially one in a "primitive" stage of development. Proper identification of an alien psychology can mean the difference between successful cultural exchange and hopelessly damaged relationships.

The problem is that "primitive" is an ambiguous

term. We think we know what it means on earth—it refers to a nonliterate culture without urban centers. The notion works fairly well here, but what does it mean when applied to beings on another planet? Knowing nothing about them while trying to fit them into a ready-made category derived from a total sample of one planet would be a gross mistake.

Proceeding from the absence of apparent signs of extraterrestrial civilizations and the inherent weakness of the "anthropomorphic hypothesis" (God created Man in His own image), S. Lem advanced an interesting hypothesis which maintains that the "nontechnological" evolution is characteristic for most existing extraterrestrial societies. According to him, the current "energetic" phase, including the expansion into outer space, constitutes only a brief period in the life of a civilization.[49]

How does one go about setting up an expedition designed to make first contact with an extraterrestrial culture? Obviously, it's too big for a one-man job, but the other extreme is equally impossible—not everyone with an interest in the problem can go. Unleashing a horde of investigators upon what seems a relatively simple culture would not only guarantee that no one would accomplish anything, but would be a sure way of disrupting local cultural patterns.

In *Unearthly Neighbors* Chad Oliver highlights the methods of ethnographic field work by imagining their application to a humanoid race on another world. For such an expedition, he chose a cultural anthropologist, a physical anthropologist, a psychologist, a linguist and an archeologist.

It's frightening to realize how ignorant we are, and how thoroughly conditioned by our own limited experiences. Stories and learned speculations about life on other planets always seem to emphasize the strange and exotic qualities of the alien worlds themselves, but the life-forms that exist against these dramatic backdrops all live like earthmen, no matter how odd their appearance may be. (Or else they live like social insects, which amounts to the same thing.) All the caterpillars and octopi and reptiles and frogs have social systems just like Vikings or the Kwakiutls or the Zulus. Nobody seems to have realized that a culture too may be alien, more alien than any planet of bubbling lead. You can walk right up to something that looks like a man—and *is* a man—and not know him at all, or anything about him.[50]

Conclusion

Direct contact is not a static, one-dimensional problem. It can take place on earth, in space or on an alien planet. As Chad Oliver has demonstrated, the more pragmatic aspects of anthropology can be applied to those entities resembling primates and in terrestroid environments. But on a larger scale, it would no longer be anthropology as we know it. All aspects of the social sciences would play important roles in the event of contact between human and extraterrestrial. No one body of knowledge is capable of studying an entire culture; astronomy, biology, the earth sciences—these sciences and others also greatly influence the evolution of life and culture. The result would be a synthesis of disciplines of both the social and physical sciences best termed xenology.

Notes

1. L. E. Orgel, *The Origins of Life* (New York: Wiley and Sons, 1973), pp. 124–129.
2. V. A. Firsoff, *Life Beyond the Earth* (New York: Basic Books, 1963), p. 91.

3. James Sutherland, "Life at a Distance," *Vertex*, December 1973, pp. 53–57, 79.

4. Sir Spencer Jones, "Origin of Life," in *New Biology*, no. 16, ed. M. L. Johnson and Michael Abercromby (Hammondworth: Penguin Books, 1954).

5. J. B. S. Haldane, *Life on Other Worlds* (London: English University Press, 1954).

6. Harlow Shapley, *Climatic Change: Evidence, Causes and Effects* (Cambridge, Mass.: Harvard University Press, 1954), p. 7.

7. Carl Sagan and I. S. Shklovskii, *Intelligent Life in the Universe* (New York: Dell, 1966), pp. 428–430.

8. Samuil Kaplan, *Extraterrestrial Civilizations: Problems of Interstellar Communication* (Jerusalem: Israel Program for Scientific Translations, 1971), p. 243.

9. Ibid., p. 6.

10. Ibid., p. 76.

11. Sagan and Shklovskii, op. cit., p. 409.

12. Gregory Benford, "Supercivilizations," *Vertex*, October 1973, pp. 49–53, 95; Kaplan, op. cit.; Sagan and Shklovskii, op. cit.

13. Sagan and Shklovskii, op. cit., p. 421.

14. Kaplan, op. cit., p. 197.

15. Sagan and Shklovskii, op. cit., pp. 428–430; Sneath, pp. 127–130, 200–201.

16. Sagan and Shklovskii, op. cit., p. 434.

17. Ibid., p. 436.

18. P. H. A. Sneath, *Planets and Life* (New York: Funk and Wagnalls, 1970), p 120.

19. Kaplan, op. cit., p. 14.

20. N. S. Kardashev, "Information Transmission by Extraterrestrial Civilizations," *Astronomie Zhurnal* 141 (1964): 282.

21. Benford, op. cit.

22. Howard Waldrop, "Lunchbox," *Analog*, May 1972, pp. 52–58.

23. Benford, op. cit., p. 51.

24. Sneath, op. cit., pp. 3–39.

25. R. H. Thouless, *General and Social Psychology,* 3rd ed. (London: University Tutorial Press, 1951).
26. Kaplan, op. cit., p. 23.
27. John C. Lilly, *The Mind of the Dolphin: A Non-Human Intelligence* (New York: Avon, 1969).
28. Sneath, op. cit., p. 155.
29. Evelyn Cheesman, *Insects Indomitable* (New York: William Sloane Associates, 1953), p. 217.
30. Pat de Graw, "Pollimander's Man-Thing," *Analog,* April 1973, pp. 161–167.
31. Weston La Barre, *The Human Animal* (Chicago: University of Chicago Press, 1954).
32. Marshall McLuhan, *The Gutenberg Galaxy* (Toronto: University of Toronto Press, 1962).
33. Sneath, op. cit., p. 155.
34. La Barre, op. cit.
35. Sneath, op. cit., p. 151.
36. James Strong, *Flight to the Stars* (New York: Hart, 1965), p. 5.
37. Sagan and Shklovskii, op. cit.
38. Ibid., p. 453.
39. Ibid., p. 463.
40. James White, *All Judgment Fled* (New York: Ballantine, 1970).
41. Perry Chapdelaine, "Culture Shock," *Analog,* May 1971, pp. 113–135.
42. Sneath, op. cit.
43. Murray Leinster, "First Contact," in *The Science Fiction Hall of Fame,* Vol. 1, ed. Robert Silverberg (New York: Doubleday, 1970), pp. 250–278.
44. Hal Clement, *Mission of Gravity* (New York: Doubleday, 1954); Hal Clement, "Starlight," *Analog,* June–August 1970.
45. Christopher Anvil, "Experts in the Field," *Analog,* May 1967, pp. 49–66.
46. Sneath, op. cit., pp. 148–150.
47. James White, *Major Operation* (New York: Ballantine, 1971), pp. 105–106.

48. Ibid.

49. Kaplan, pp. 244–246.

50. Chad Oliver, *Unearthly Neighbors* (New York: Ballantine, 1960) , pp. 33–34.

Further Readings

Anderson, Poul. "The Ancient Gods," *Analog*, June–July 1966.

————. "People of the Wind," *Analog*, February–May 1973.

————. "Wings of Victory," *Analog*, April 1972, pp. 60–75.

Ash, Paul. "Minds Meet," *Analog*, February 1966, pp. 137–145.

Benford, Gregory, and David Book. "Is Anybody Out There?" *Amazing*, March 1970, pp. 114–119.

Bova, Ben. "Galactic Geopolitics," *Analog*, January 1972, pp. 51–62.

Chapdelaine, Perry. "Initial Contact," *Analog*, May 1969, pp. 143–162.

Clarke, Arthur C. *Childhood's End* (New York: Ballantine, 1960) .

————. *The Promise of Space* (New York: Harper and Row, 1968) .

Cook, Rick. "Life as We Don't Know It," *Analog*, November 1970, pp. 39–59.

Fisher, Gene. "Stimulus-Reward Situation," *Analog*, August 1973, pp. 65–77.

Green, Joseph. "One Man Game," *Analog*, February 1972, pp. 112–124.

Kessing, Roger, and Felix Keesing. *New Perspectives in Cultural Anthropology* (New York: Holt, Rinehart and Winston, 1971) .

Kleine, Walter L. "Cultural Interference," *Analog,* April 1969, pp. 87–107.

Le Guin, Ursula K. *The Left Hand of Darkness* (New York: Ace, 1969) .

Leinster, Murray. "The Aliens," *Astounding Science Fiction,* August 1959, pp. 8–40.

Lowell, Percival. *Mars as the Abode of Life* (New York: n.p., 1909) .

"Message from Mankind," *Time,* March 6, 1972, pp. 68, 72.

Mucchielli, Roger. *Introduction to Structural Psychology* (New York: Funk and Wagnalls, 1970) .

Reynolds, Mack. "The Enemy Within," *Analog,* April 1967, pp. 57–70.

Schmidt, Stanley. ". . . And Comfort to the Enemy," *Analog,* July 1969, pp. 6–37.

White, James. *The Aliens Among Us* (New York: Ballantine, 1969) .

———. *Hospital Station* (New York: Ballantine, 1962) .

———. *The Secret Visitors* (New York: Ace, 1957) .

———. *Star Surgeon* (New York: Ballantine, 1963) .

Wilhelm, John. "Is There Life on Mars—or Beyond?" *Time,* December 13, 1971, pp. 50–58.

Wodhams, Jack. "The Visitors," *Analog,* September 1969, pp. 140–157.

Young, Robert F. "Genesis 500," *Analog,* February 1972, pp. 50–64.

B.
The Moral Obligations of Anthropology

BARBRA D. MOSKOWITZ

If and when we encounter new life forms in inter-planetary travel, anthropology will be outdated as a term, though certainly not as a useful social science. The definition of the word anthropology as "the study of man" must be expanded to include the study of all intelligent life forms. In this context "intelligent" is not used as a value judgment but refers to the organization of life forms that function

* For the purpose of this paper I have chosen to use the word "interplanetary" in place of "extraterrestrial." The latter is an ethnocentric term, and may prove psychologically as well as physically harmful in the future.

I would also like to acknowledge the assistance of Professors Richard O. Clemmer and Fred T. Plog III in the preparation of this paper.

as cultures—cultures probably unfamiliar to us in form but not in content. It is to this content that anthropologists must address themselves. By offering clues, comments and interpretations expressed in our own culture's terms of understanding, anthropologists can play an important part in exploring space.

Historically, exploration has been disastrous for the cultures encountered along the way. The reactions of the explorers to people only slightly different from themselves in physical appearance was appalling, and "savages" were repeatedly sent back to Europe for show alongside animals and plants found in the "discovered" lands. They were generally given little respect as intelligent beings living in viable cultures.

Our confrontations, on the other hand, will most likely be with life forms as yet unfamiliar to us—physically, biologically, perhaps even psychologically. The realization of past mistakes as well as the facts and interpretations gleaned from previous cross-cultural encounters are necessary to command in ourselves attitudes of respect for the lives and life styles of interplanetary inhabitants.

In considering the existence of interplanetary cultures, it is naïve to imagine all cultures as "less" civilized than ours. When circumstances arise so that we are in contact with "more" advanced beings, our entire orientation to the contact situation will change. Our goal at that point will be to live through the experience. Choice in action will be ours only insofar as it is granted to us. Our studies and researches will be limited, if existing at all. It is quite possible a more advanced culture will not react belligerently to us, but this is not the point. In the definition of more or less advanced, it is the potential for destruction that is important. Anyone operating under the weight of possible destruction is not free.

Another possibility is that a culture might have known of our existence previous to contact. It may have come to the conclusion that we are not worth bothering about, and in such a case would be bored and uninterested in any sort of contact situation with us. Or we could be so far behind that communication would not even be possible. The only reaction to these situations would be to travel on, hoping that some time in the more distant future, reasonable and mutual communication might be set up.

Some of the more fantastic ideas on interplanetary life involve the attitude and psychology of the human being. We might interpret an extremely advanced civilization as "godlike," or we earthlings might be mere marionettes, or experimental subjects of an interplanetary civilization. With just a slightly greater stretch of the imagination, we might find that we actually have been created by other beings.

Although there is no reason to assume that a highly advanced culture would welcome our inquisitiveness, we might hope that at least some of them will accept us without threatening us. In such situations a balance might be found between their generosity and our curiosity, where we might gain quite a bit of knowledge at the expense of no one. This would certainly be a scientist's heaven, and the contacted culture would sooner or later probably have to set up immigration laws. Further consideration of our attitudes in situations where we are the "less" advanced culture are more theoretical than other discussions, as decision-making on more than a very basic level will—at least at first—be out of the anthropologists' field of study or influence. Therefore I will limit my discussion to contact situations with beings which are more or less at the same level as *Homo sapiens.*

With insights from the social sciences, we should be able to view all functioning societies as complex and intelligent, as viable adaptations to a given time and place, and deal with these cultures and their members accordingly.

However, such acceptance is slow. For the natives of the Americas as well as other continents, the situation remains bleak. Americans are not innocent in the planned genocide of Brazilian tribes: the road being built through the Amazon is financed in part by U.S. dollars, and has been used as an excuse to murder thousands of native people. Between 1950 and 1968 the 30,000 members of the Cacaas Novas tribe have become 400. The Bocas Negras were destroyed by flu caught by contact with Westerners; those not dead from the flu were shot to death. Only a handful escaped.

There are, however, signs that positive changes are increasing in number. In 1973 the Menominee Indians won reinstatement as a federally recognized tribe after a long battle that began in 1954, when President Eisenhower signed the tribe into termination. At that time they had been fairly prosperous, learning the "ways of the white man," and running tribal businesses, with money in the tribal treasury. Termination meant the end of federally supported schools, health care, and other projects. It broke the land into privately owned lots, an unfamiliar concept to tribal mores. It became impossible for the tribe to support the previously government-funded projects, which led to the sale of most of the land to resort developers and lumber companies for a few cents an acre. Since termination, the tribe has argued and lobbied for reinstatement: in 1973 they won their

legal battle. The Northern Cheyenne also won a
recent victory over the strip-mining companies that
had threatened 60 percent of their 415,000-acre reser-
vation in Montana.

There is still, however, a long way to go: school-
teachers continue to beat children for speaking their
native language, or for following their traditional
religion; three hundred and seventy-one treaties
broken by the U.S. government are ignored; Indian
lands are taken over for industrialization, and tribes
are pushed onto smaller and smaller reservations.
The Havasupai are currently fighting to retain their
land near the Grand Canyon, wanted for expansion
of the Grand Canyon National Park within which
they would be prohibited from hunting, picking
herbs, planting, using water and even from gathering firewood, all essential to their way of life.

To consider anthropology as a tool in interplanetary culture contact, it is first necessary to reassess its
role in terrestrial cross-culture contact. In the past,
anthropology has been used as a means for assimilat-
ing other cultures into the American (or Western)
mainstream, leading inevitably to the destruction of
those cultures. The Dawes Act of 1887 was partially
formulated and supported by an anthropologist,
Alice C. Fletcher, and granted tribal land in sev-
eralty to members of a tribe, rather than as one large
piece of land granted in the traditional manner to
the tribe in common. Though Fletcher thought sev-
eralty the best thing for the Indians, most Indians
felt differently. In fact, most native peoples are not
intent upon having their cultures destroyed and
becoming totally assimilated; instead, they want responsibility for their own lives and the lives of their
families and tribes along with fulfillment of treaty
obligations, treaties imposed upon them by the U.S.

government. Rather than remaining spokesmen and deciding what is best for another culture in our own view (and for our own benefit), anthropology must become a tool for communication between and among different life forms.

Another function of anthropology, at least as important as the above, is to explain our culture—our myriad of cultures to those unfamiliar with us. Too often in the past we have forgotten this side of our discipline, which has resulted in, for example, misunderstandings between native Americans and the United States in land exchange, in religion and in the basic communication of trust and honesty. We will need to apply these basic concepts of trust and honesty to our confrontations with interplanetary cultures, and to exchange concepts as well as words and goods with them.

Just as any social group or culture we meet in interplanetary travel may not be representative of other groups on its own planet, and certainly not on other planets, our scientists and social scientists in the contact situation will not be representative of our own culture. They will be a new and highly specialized subsystem of earth, which in itself is composed of various and divergent cultures and systems. The crews of interplanetary spaceships will be dependent upon earth for their entire environment— food, clothing, knowledge, even the environment itself—but will function as a separate system, appearing complete, at least on first encounter, to any interplanetarians met.

Anthropologists must be prepared to make our culture as a whole comprehensible to those who meet us. Since our main goal will be to survive in any interplanetary situation, possibly our very survival will hinge upon our ability to explain our own culture and intentions (presumably pacific) to the interplanetary peoples, in, terms understandable to them. If we are prepared to interpret foreign life styles into our own cultural concepts, we will be able to reverse the process. As Leighton stated in *The Governing of Men* (1945) and as is paraphrased by Kennard and MacGregor in 1953, "it is equally important to study the assumptions, social organization, and behavior patterns of the administrative group as well as of those administered, since the two groups constitute an interacting continuum."[1]

Misconceptions and assumptions on either side about the other are dangerous. We have no reason to expect an interplanetary culture to interpret our arrival as a peaceful venture. Indeed, throughout history, possibly because we were on the winning side, we have maintained that "might makes right," though morally we espouse the opposite, the Christian belief that "the meek shall inherit." Actually it is the use of threat and the assertion of any power that controls or destroys which makes a society "more" or "less" civilized than another.

In situations where we maintain power, our orientation and goals as anthropologists will vary. It is quite possible that our terrestrial problems of social unrest, collapsing and expanding economies, ecological disaster and political upheaval may force us into space, demanding—or seeming to demand—manipulations of contacted life forms and life styles. Such excuses as the above have led to exploration in the past, and it is reasonable to assume that they will, at least in some degree, do so again.

When England was exploring and settling new lands she found herself with overcrowded debtors' jails. To alleviate this problem, and to claim new lands for herself at the same time, she established the colonies of Georgia and Australia as debtors' colonies, places certainly more pleasant than the jails of the Old World. Today jails, houses of correction and mental institutions remain overcrowded and dismal. Moreover, a large proportion are tax-supported institutions in which the taxpayer generally has little interest. The public becomes concerned only when inhumane conditions are dramatically brought to its attention.

As long as the possibility of establishing a penal colony on a physical body in space exists, anthropologists must be concerned with the type of settlements as well as with the effect of such settlements upon possible indigenous beings. Our primary concern is not the prisoners, but the "right of discovery" possibly claimed in such an instance. This "right" can be found in Chief Justice John Marshall's decision in the case of *Johnson v. McIntosh:*

> The potentates of the world found no difficulty in convincing themselves that they made ample compensation to the inhabitants of the new world, by bestowing on them civilization and Christianity, in exchange for unlimited independence. But, as they were all in pursuit of the same object, it was necessary, in order to avoid conflicting settlements, and consequent war with each other, to establish a principle, which all should acknowledge as the law by which the right of acquisition, which they all asserted, should be regulated between themselves. This principle was, that discovery gave title to the government by whose subjects, or by whose authority it was made, against all other European governments, which title might be consummated by possession.[2]

This led the United States, as well as other nations, to maintain "that discovery gave an exclusive right to extinguish the Indian title of occupancy, either by purchase or by conquest."[3]

People now claim to be more "civilized," "advanced" and "humane" than those conquerors of the seventeenth, eighteenth and nineteenth centuries. As anthropologists it is our job to see that this is true and to see that the theory of equality generally proclaimed for individuals is upheld for cultures as well. Most important to reaching the goal of equality for cultures is the understanding of the social, political and religious relationships of alien life forms.

The misunderstanding of native cultures was aggravated by the historical fact that accompanying Manifest Destiny were industrialization and capitalism, and the "big business" concerns and concentration of money these created. Since in our society money is an indicator of power, those with money have the ability to direct the use of it. One of the few redeeming values of this authority is that it makes research grants available. Funding of some sort is essential to any science. However, anthropologists will have to be wary of economic and political pressures and the solutions they may be asked to validate. Among the foremost dangers is the development of new sources of natural resources found on inhabited planets. For the sake of earth's "economy," and undoubtably to the detriment of the society involved, there will be immediate demands for the rights to have minerals, oil, etc. Historically it has always happened that if honest trading does not work fast enough, lying, cheating and stealing follow in short order.

Consider, for example, the Peabody Coal Company's strip-mining now taking place on Black Mesa, Arizona, a mesa that is sacred to both the Hopi and Navajo peoples. Salvage archeology is being conducted in the area. But what are being "salvaged" are the things important to *our* culture, *our* study, artifacts in *our* eyes, while what is of the utmost importance to the people we are "salvaging" is being destroyed—that is, the sacredness of Black Mesa itself.

Suppose that rich underground oil fields were discovered in Jerusalem and the Israeli government agreed to sell this oil to an American oil developer at low prices; the company on its part agreed to leave just enough time before it began drilling to allow archeologists to remove the Western Wall and the Dome of the Rock, stone by stone, from top to foundation. An outcry would be raised across the world, supported by Arabs and Jews alike, for it is not the stones and gold that are important, but the symbolic and spatial location itself.

This is improbable in Jerusalem, but it is precisely what is happening with coal on Black Mesa, oil in Alaska, park land in Arizona, and tin in Brazil. The same situation could conceivably develop on a distant planet inhabited by an unfamiliar life form. It is anthropologists who must explain and defend the concept of what is sacred to a people and similar relationships between intelligent beings and their systems.

In the rush to develop as an industrialized culture Western man has depleted the earth's natural resources, creating shortages of nearly all raw materials. Men and women may be forced into space either to abandon a dying planet or to search for new sources of raw materials to maintain its present life style. However, in considering the technical development of such resources, men must be concerned with the criteria to be used in exploiting these resources on inhabited planets.

The priority consideration is the attitude contacted life forms have toward their natural environment. Anthropologists must make sure that the prospect of use of *their* land is presented to them fairly and in terms they can verbally and conceptually understand, and that their decision is upheld. A contacted culture's relationship to its own environment is something basic, dependent and important to that culture. If and when a contacted culture denies us the right to develop its resources, politicians, as foreign policy makers, will have to define a course of action. Anthropologists, as cultural interpreters, must present the views of the culture involved.

The extent to which anthropologists present and defend these views depends on the position they decide to take. They can permit politicians influenced by economy, industry and the acquisition of power to formulate the directives in dealing with interplanetary cultures. In that case, anthropologists supply reasons, excuses and rationalizations for the treatment, whether good or bad, of newly "discovered" life forms. Or, they can demand a voice—along with those in other sciences and social sciences—in the formulation of those political directives, and in keeping the goals of the exploration of space in balance with the demands of the interplanetary peoples.

One of our main goals in space travel has always been the acquisition of knowledge—that is, the accumulation of facts. We must not only accumulate facts, but use them to improve our own systems. Working from the outside in, in a kind of reverse ethnocentricity, anthropology must help present new cultures and new life styles not as threats to our existing system, but as ideas and alternatives that might be adapted to improve our own culture. Consider, for instance, the American Indians' respect for nonhuman as well as human life. If we had learned from them rather than imposing our own beliefs upon them, there would be more species of animals and plants in existence and many fewer in danger of extinction. The eagle, for example, is held in reverence by many Indian nations. We have shot them nearly out of existence, and as a result, have established an imbalance in nature with respect to the species upon which the eagle preys. Interplanetary culture contact will give us the unique opportunity to see our culture from the point of view of nonhuman intelligent beings.

Prior to any actual interplanetary culture contact we must decide on specific directives. It is assumed that once contact is established, we will modify and develop the original directives to meet the demands and needs of each particular circumstance, but to begin, two very general directives can be applied: the Theory of Reciprocity, and the Doctrine of Free Choice. The first is already well-known under another name: the Golden Rule. Here, it is used not as a religious doctrine but as a political and social ethic. The Theory of Reciprocity is simply this: respect others in the manner you yourself would want to be respected, and act accordingly. We must conduct ourselves at all times with the possibility well in mind that although at the time we may be in contact with a culture "less" civilized than ours, there may come the time when we meet a culture able to threaten us with destruction. The considerations we would expect them to grant us should be the same ones we grant others.

All cultures, at every stage of development, represent an intelligent life form's adaptation, use and manipulation of its environment. The decisions made by any society or organization of individuals, however organized, the knowledge gained from these decisions, and their further and continual adaptations must be seen as dynamic factors, as the continuing evolution of that culture. Any fixed point at which we come into contact with a culture is only that—a point in a continuum of change and growth, a continuum which does not necessarily denote a linear progression. There is the probability that in our space explorations we will meet many cultures at various points of development. It is imperative that we not judge a culture by an arbitrarily fixed point —in this case, that of contact. In fact, we must not judge a culture at all. It is too presumptuous to consider our own culture as a model of "civilization" or even in an "advanced" position on the continuum, for we also are ever-changing.

At one time it was believed that the earth was the center of the universe and that the sun, moon, stars and planets revolved around us. We now know that only the moon does so, that the sun and stars travel around a distant center, and that this center revolves around another, ad infinitum. Yet we still tend to think of ourselves or of our own culture as the "center," as the "best of all possible civilizations." It is essential, however, that the subsystems of our culture that come into contact with other cultures, the specialized social groupings in spaceships, reject all ethnocentric theories and doctrines.

It might prove quite dangerous to hold to our practices of ethnocentrism in space travel. In the past, when we have held the Christian ethic and practice of conversion on a pedestal in our dealings with other cultures, we have assumed that Christianity was the righteous religion and that all others were "heathen," "savage," "primitive" and "barbaric" practices evoked by the devil. We forcibly and threateningly have converted natives all over the world to our religious beliefs, with varying degrees of success. There is a great danger that we might continue this practice, and after establishing contact with an alien culture, send missionaries into outer space to "convert and civilize" them—with, of course, all good intentions on our part. In such a case there is bound to be a forced breaking of sacred and often secret religious rituals and artifacts and a denouncement of native beliefs.

The Theory of Reciprocity must be coupled with the Doctrine of Free Choice. We can repeat over and over that we will deal with interplanetary life forms equitably, but that does not mean that we will do so. Look, for instance, at history and the words in treaties made between the United States and the Indians: "The utmost good faith shall always be observed towards the Indians: their land and property shall never be taken from them without their consent; and in their property, rights and liberty, they shall never be invaded or disturbed . . ." (from an Ordinance for the Government of the Territory of the United States northwest of the river Ohio, 1789). "The United States will forever secure and guaranty to them, and their heirs or successors, the country so exchanged with them . . ." (the Indian Removal Act, 1830). And another example, from an address by President George Washington in 1790 to the Six Nations: "The General Government will never consent to your being defrauded, but will protect you in all your just rights." And in 1926 the Department of State before the American and British Claims Arbitration: "Under that system the Indians residing within the United States are so far independent that they live under their own customs and not under the laws of the United States . . . and whenever those boundaries are varied . . . they receive from the United States ample compensation for every right they have to the lands ceded by them."[4] The list goes on and on, promising in nearly every treaty or agreement made with any tribe, or in laws pertaining to Indians in general, that the United States will protect the right of the Indians to their lands and properties. But history shows that the government reneged on these promises.

The basic problem with the treaties made by the United States and Indian nations is that the United States forced its laws, standards and psychology upon them. In the future it is essential to practice the Doctrine of Free Choice, the principle that any life form, any cultural entity or part of a cultural entity as defined in its own terms has the right to accept or reject, under no duress, temptation or threat, any proposed changes we might offer in their life styles or in the management of their property. We must accept *in toto* all political, ecological, social, religious, scientific, and moral decisions reached by the members of each particular culture.

Although a culture might not be organized precisely into traditional categories (religion, art, economy, politics, kinship), these can provide a starting point for understanding how each culture we contact functions. Religion is how the unknown is transformed into a known and how this is communicated among individuals of the society. The communications can be through myth, legend, oral tradition, laws, taboos, written texts, rites and rituals. We must remember, however, that an action or ritual that we recognize as religion may be based on something other than what we define as "unseen" or "unknown," and a people's explanations and actions must be taken into consideration. If, for instance, what we mistakenly interpret as superstitious is actually based on a concrete reality, what we had considered as religious could have been political, social or scientific. What classifies an individual's action or belief in any culture is that individual's attitude toward it, and the society's reaction to that—*not* the anthropologist's projections.

Art (literature, design, painting, sculpture) is another way of communicating. We will come into contact with forms of art we never could have imagined. The particular culture's attitude toward the work—is it purely functional or is it an accessory and a personal expression of some sort?—can be an invaluable source of information. Another part of art is the decoration of any component of the technology, such as garments, tools or buildings. Anything not essential to the actual function of a technological object, or anything expressing individualism, can be considered an artistic characterization. The amount of décor existing can be a valuable clue to the amount of individualism accepted and encouraged in a given society.

The economic system can be broken down into three areas: subsistence, technology and trade. With unfamiliar life forms we are bound to encounter unfamiliar subsistence patterns. A life form may have a different energy base than we know. It may take energy directly from its sun, for instance. Its technology, or how it acquires its subsistence needs, will be harder to pinpoint. Trade, if it exists at all, will also be different. Anthropologists must search for the patterns as well as the manifestations of them, as they may not be at all apparent.

The political system, its laws, leaders, power and

application of power, also must be accepted through the culture's viewpoint. Too often in the past we have forced our own political system upon contacted cultures. Partially to ease government dealings with various Indian tribes, we forced a Western system of elected government on all Indian tribes, regardless of the existing type of leadership they had. We must accept the leadership of those designated to deal with us. Certainly we can take into consideration the social status of the leader, be it an individual or a system. Whether a war leader or a religious leader (if these positions apply) negotiates with us will certainly make a difference in the way the culture reacts to us and the way we react to it. If they consider us a threat they will send a war leader; if they consider us a supernatural occurrence, they will send a religious leader. The confrontation would contribute to our understanding of the culture and its psychology.

The study of kinship systems has been based quite logically on our knowledge of the human reproductive experience. It can be expected that unfamiliar life forms will have totally different methods of reproduction and that their kinship systems will reflect this. What appears to be a parent/child relationship might be a social relationship instead, and parent/child concepts as we know them might not exist at all. Imagine the reversal in kinship relationships if in an interplanetary culture children are born with total knowledge and lose it as they age rather than gain it. The dependency situation might be opposite to our own.

The immense number of possible variations of cultural manifestations will leave us in constantly uncertain positions in contact situations. How to act, when to act, how to make ourselves understood and how to understand interplanetary peoples are only the basic problems that will face the spaceship crews. In initial contact the anthropologists' role will be to as quickly and as accurately and un-ethnocentrically as possible assess the contacted culture, an assessment that should be used *only* for immediate action and communication. Cultural description and interpretation requires much research and involves long periods of intense field work, hopefully possible in the contact situation. Only after continuous in-depth research can anthropologists assert their necessary role as cultural interpreters, but not as spokesmen for interplanetary cultures. To assume that once one understands the functionings of a society, even the psychology of its individuals, one can present the feelings and views of that culture better than the constituents is incorrect. Anthropologists must listen to the demands, thoughts and requests of the culture, make objective sense of them, and communicate them to earth. As interpreters, anthropologists will not only be interpreting language, but also the manifestations of culture, the decision-making processes and the continuum of change upon which a particular culture finds itself.

To act as such interpreters, and to incorporate into such action the Theory of Reciprocity and the Doctrine of Free Choice, anthropology will be acting as a "moral science." There can be no repetition of the past. Under the guise of "goodness" and morality thousands of people were killed in the Crusades, thousands in the settling of the Americas by the Spanish conquistadores, thousands in the industrial advancement of Western countries in America, South America, Africa and Australia. Anthropology now gives us the perspective and ability to compare all cultures, no matter how "primitive" or "advanced," to our own, *without* making value judgments.

As Jennings C. Wise stated in 1925 in "A Plea for the Indian Citizens of the United States":

> Let the historian speak the truth, not out of a bitterness of heart, not with a vindictiveness designed merely to brand with unanswerable accusations a people who profess to be repentant, but to make the world so deeply conscious of their sin that others may pause upon the threshold of conquest to ask God what is the obligation which civilization imposes along with the right of preemption that may be claimed in its name.[5]

It is this obligation, this moral obligation, that must be considered in interplanetary contact.

As cultural interpreters, anthropologists can and should be an integral part of space exploration. They can contribute greatly to the accumulation and expansion of knowledge, as well as to the safeguarding of other life forms.

Notes

1. Edward A. Kennard and Gordon MacGregor, "Applied Anthropology in Government: United States," in *Anthropology Today*, ed. A. L. Kroeber (Chicago: University of Chicago Press, 1953) , p. 837.
2. As quoted in Vine Deloria, Jr., *Of Utmost Good Faith* (New York: Bantam, 1972) , p. 43.
3. Ibid.
4. Ibid., pp. 62, 211, 218.
5. Ibid., p. 302.

4.
Extraterrestrial Communities— Cultural, Legal, Political and Ethical Considerations

PHILIP SINGER

CARL R. VANN

A continuous thread runs through all considerations of extraterrestrial communities,[1] from Plato's B.C. ideal Republic to Noyes' nineteenth-century Oneida utopian community—that of authoritarianism[2] supported by a profound unawareness of cultural behavior. Indeed, the current metaphors used for extraterrestrial communities—vivarium, test tubes, satellites—connote this authoritarianism.

The assumption here is that the new communities can combine technology (bubble, dome, etc.) and science (research) to produce significant data. This assumption might be valid in engineering, biology and medicine, but does it have anything to contribute to the social sciences, most specifically anthropology? We are not sure.

Clearly any extraterrestrial community will be a planned experiment, just as all space exploration, manned or unmanned, has been to date. Thus culture becomes part of the input variables, instead of the matrix itself. It is important, in any consideration of extraterrestrial communities that this distinction between culture as a given, which anthropologists study, and culture as a manipulatable variable, be kept in mind. Otherwise, we shall confuse "community" with "experiment."

By experimentation, anthropologists generally mean use of the comparative method. Nadel quotes Talcott Parsons as saying: "The experiment is . . . nothing but the comparative method where the cases to be compared are produced to order and under controlled conditions."[3] As Lewis says, basing his comments on Nadel and Parsons: "Since, in the study of culture, we cannot as a rule produce the artificial induction of variations under controlled conditions, we do the next best thing and study variations as they occur over time and compare and correlate."[4] This basic question that is raised for anthropologists and for all students of culture and society is, What is the scientific-experimental approach to the study of cultural behavior? Haring says:

> Human social behavior . . . is cultural (i.e., learned from other people). New behavior may be learned by anyone at any time and substituted for former behavior. These facts invalidate laboratory studies of social phenomena. To suggest a crude example, it would be ludicrous to select two individuals of opposite sex, place them in a laboratory, and instruct them to fall in love so that the process might be analyzed. Even should the subjects thereupon "fall in love" there is slight reason to expect that two other subjects would duplicate the first experiment. Failure to develop crucial controlled experiments, however, does not mean that social and psychological hypotheses cannot be tested. A hypothesis that holds under varying and contrasted cultural conditions is validated to that extent. If in a contrasting cultural milieu, the hypothesis no longer fits the facts, it is not valid universally, whatever its local applicability.[5]

What can extraterrestrial communities enclosed in vivarium technology contribute to the study of man? We will learn nothing about the origin and development of culture, that most insoluble of anthropological questions, for the community may be a test tube, a closed ecological system, but with the entrance of the first settlers, it will be contaminated, for every person in it will already be a "culture bearer." There is no such thing as a "test-tube culture," since "pure persons" in a "pure culture" do not exist. Presumably an extraterrestrial community will have to be based on a "balanced life system" as understood by biologists, because of the finiteness of resources. However, the concept of "space-ship earth" as an (im-) balanced life system contains much greater possibilities for modification because of the factor of size (some would say indefinite modification) than does an artificial spaceship community with finite self-sufficient life-support systems. It seems to us that the entire concept of a balanced life system which is inherent in extraterrestrial communities implies a research approach in terms of dynamic equivalence, as understood in the physical or "hard" sciences, because predictability would of necessity be required to maintain the limited system. Human beings may be conceived of as "systems" whose internal response mechanisms are indefinitely modifiable and subject to capricious changes. Such changes are not compatible with a limited balanced life system. But "all significantly related events of social phenomena are dynamically inequivalent."[6] The notion of "reinforcement," "control," "measuring of responses," etc., all share a mechanistic or energistic affinity with the first law of thermodynamics, the conservation of energy. If there is anything that the comparative study of cultures has demonstrated, it is that the response of living organisms bears no predictable relation to its stimulus. However, if the extraterrestrial community is to survive, it would of necessity have to be based on Skinnerian reinforcement schedules eliminating all possibility for unpredictable change, or people simply changing their minds. The only kind of cultural behavior that could be meaningfully gathered would be a description of persons behaving with certain frequencies for certain kinds of behavior whose categories have most likely been preselected by the behavioral scientist.[7] The fact is that the "stimulus history" of an organism exhibiting cultural behavior defies measurement.

Given the "indefinite range of individual modifiability in pattern of response,"[8] or, more simply put, that people change their minds, a "balanced life system" in a bubble or in a "black box" would not provide the cultural anthropologist with a very meaningful experiment. With what would it, or could it be compared? It could not be considered another culture, because it is already part of a larger culture. The cultural internal mechanisms of the homeostatic extraterrestrial community could not be discerned because of the modifiability of cultural behavior. Culturally, at least, the "community" would be unbalanced and internally unstable. Of course, the psychocentrist might argue that an extraterrestrial community balanced life system or a black box system would only give the appearance of being internally unstable, if one was ignorant of important internal mechanisms and thus unable to take into account many critical factors. One could argue that the basic datum would be the rate at which certain preselected responses were emitted, and that this type of measurement could be done automatically, as indeed is the case in measurements involving the astronauts. One would have to carefully select the extraterrestrial community inhabitants and get good base-line data—verbal, motor, social, sexual, etc. If one accepts this view of the dynamic inequivalence of cultural behavior and the indefinite modifiability of human response, then all discussion concerning cultural "universals" becomes an arid exercise in classification. Indeed, most anthropologists would agree that there are as many expressions of the universals of culture, such as family, language, housing, technology, art, as there are cultures that have been studied. Therefore, extraterrestrial community experiments that try to predict cultural behavior, instead of just describing it, will fail. That the community would perforce be a closed ecological system does not simplify the problem, and the modifiability of cultural behavior within that system causes one to question the entire concept of a state of equilibrium.

Stewart quotes Forde as saying:

> Neither the world distributions of the various econ-
> omies nor their development and relative importance
> among the particular peoples, can be regarded as
> simple functions of physical conditions and natural
> resources. Between the physical environment and hu-
> man activity there is always a middle term, a collec-
> tion of specific objectives and values, a body of knowl-
> edge and belief: in other words, a cultural pattern.
> That the culture itself is not static, that it is adaptable
> and modifiable in relation to the physical conditions,
> must not be allowed to obscure the fact that adapta-
> tion proceeds by discoveries and inventions which are
> themselves in no sense inevitable and which are in
> any individual community, nearly all of them acquisi-
> tions or impositions from without. The peoples of
> whole continents have failed to make discoveries that
> might at first blush seem obvious. Equally important
> are the restrictions placed by social patterns and re-
> ligious concepts on the utilization of certain resources
> or on adaptations to physical conditions.
>
> The habitat at one and the same time circumscribes
> and affords scope for cultural development in relation
> to the pre-existing equipment and tendency of a par-
> ticular society, and to any new concepts and equip-
> ment that may reach it from without.
>
> But if geographical determinism fails to account for
> the existence and distribution of economies, economic
> determinism is equally inadequate in accounting for
> the social and political organizations, the religious be-
> liefs and the psychological attitudes which may be
> found in the cultures based on those economies. In-
> deed, the economy may owe as much to the social and
> ritual pattern as does the character of society to the
> economy. The possession of particular methods of
> hunting or cultivating, of certain cultivated plants or
> domestic animals, in no way defines the pattern of

society. Again, there is interaction and on a new plane . . . The tenure and transmission of land and other property, the development and relations of social classes, the nature of government, the religious and ceremonial life—all these are parts of a social superstructure, the development of which is conditioned not only by the foundations of habitat and economy, but by complex interactions within its own fabric and by external contacts, often largely indifferent to both the physical background and to the basic economy alike.[9]

The present state of scientific knowledge is always incomplete. However, we wonder whether the "laboratory experiment" type of extraterrestrial community would offer new data to the study of human behavior.

Nevertheless, perhaps such an undertaking is justified as an interdisciplinary effort which would demonstrate to scientists other than anthropologists that ecological analogies, biological models, physical-science metaphors and ethological speculations are not adequate to describe or explain cultural behavior. The political and economic aspects of human experimentation within a vivarium (for that is what an extraterrestrial community would be) are directly related to man's theoretical and practical attempts to construct ideal societies, without artificial enclosures or the balancing of the oxygen-nitrogen cycle. The utopian tradition from Plato to our own time hypothesizes that the external manipulation of the environment can bring society to a static perfection with fixed and clear human relationships and functions. No matter how idealistic or seemingly desirable such proposals may be, to succeed there must be an element of coercion. The basic assumption underlying the utopian tradition is that men agree on the goals of their society and believe in their ability to act in ways that will conform with those goals. We have no empirical reason to believe that man's nature is such that this agreement is possible, or that the perceptions of society's goals are similar enough to allow for an on-going system in which all necessary functions are adequately performed. Life within a vivarium or any society involves an internal authority structure, rule making and rule enforcing procedures and the external problem of relationships with possible manipulations from those outside of the system. John Donne put it nicely: "No man is an *Island,* entire of it selfe; every man is a peece of the Continent."

In any extraterrestrial community, as space programs have demonstrated, it appears that the only type of authority structure which can be used must involve the military model. Any other attempt to structure a system for choosing leaders, devising, interpreting and enforcing rules and pushing switches would conflict with the basic technological assumptions of the community and could easily lead to chaos. However, the military model, a highly interdependent structure, is not without similar risks. Structuring a system with hierarchical command does not in itself guarantee compliance, and the entire extraterrestrial community experiment could come to a rapid halt in the event of any appreciable amount of deviant (cultural) behavior. The cultural-political system of the vivarium community would thus be based on some type of group theory of organization involving the subordination of the individual to the collective good. The same would hold true of the vivarium's economic relationships. Given a highly structured, highly interdependent closed system, all functions performed are vital to the continuation of life processes. It would therefore be difficult to devise or utilize a system of rewards and services that emphasizes any special role or remuneration for the individual based on his personal strivings. Some type of economic relationship more related to the Marxist concept of "from each according to his ability, to each according to his needs" would be necessary. Any other type of structure would run the risk of breakdown because of conflict due to inequalities in the treatment of individuals. The use of a "community" vivarium for human experimentation raises a variety of legal and constitutional questions, at least for Americans.

We are, however, dealing with extremely hypothetical situations, no matter how specific the balanced life space proposal may be. Since our legal system only considers real—as opposed to hypothetical—cases or controversies, the possibility always exists, no matter how carefully something may be planned in advance, of a dynamically inequivalent nuance occurring in which some legal right may be involved.

The rights of citizens under our Constitution extend primarily to protections against the exercise of arbitrary governmental authority. Although it is certainly true that the notion of what is public and private has undergone considerable change over the past century, the problems of research on human beings fall primarily into the sphere of private human relationships. There are two categories where the government is significantly involved. There is the role of the government as a sponsoring research agent which involves both institutional liability and ethical considerations. This type of research sponsorship involves private citizens supported in their work by government grants, civilian government officials who engage in research functions as part of their jobs, and persons in military service who may be involved in research in any of several roles. Our constitutional system places the military in a special category and to a very large extent this may govern much of extraterrestrial human participant observation (experimentation). The Constitution specifically empowers Congress:

> To provide for organizing, arming, and disciplining the militia, and for governing such Part of them as may be employed in the Service of the United States, reserving to the State respectively, the Appointment of the officers, and the Authority of training the militia according to the discipline prescribed by Congress.
> (Article 1, Section 8)

In addition to this rather broad grant to set up special rules for the governance of the military, there is a provision in the Fifth Amendment to the Constitution which places the military in a special realm with respect to rights in the criminal process. It reads that:

> No person shall be held to answer for a capital, or otherwise infamous crime, unless on a presentment or indictment of a grand jury, except in cases arising in the land or naval forces, or in the militia, when in actual service in time of war or public danger; nor shall any person be subject for the same offense to be twice put in jeopardy of life or limb, nor shall be compelled in any criminal case to be a witness against himself, nor be deprived of life, liberty, or property, without due process of law; nor shall private property be taken for public use, without just compensation.

The exception of the military in the Fifth Amendment authorizes "the trial by court-martial of the members of the armed forces for all that class of crimes which under the Fifth and Sixth Amendments might otherwise be deemed triable in the civil courts" (*Ex Parte* Quirin, 317 U.S. 1 at 43, 1942). Clearly, then, the rights of persons in the military relative to human research communities would be considerably more flexible than civilians and that projects involving rigid authority and role structures could probably be attempted more easily with them. We do know that Army regulations require the consent of soldiers to medical procedures designed to enable them to properly perform their military duties.[10] The same authority concludes that military personnel are prohibited from consenting to any experiment that might interfere with the proper performance of their military duties.

In general, however, the military model is based upon the authoritative hierarchical receipt of approval for an individual to volunteer for an experiment. To enhance this goal in the area of human research, the Army issued Special Regulation 70–30–1 in 1949, "Research in Human Resources and Military Psychology." Although the exact limits of the use of this regulation are not known, it would seem that projects of the extraterrestrial type could best be undertaken within the military establishment. It should be added, however, that this would govern the individual service member only and not his family.

We would expect that the most important cultural-legal problems relating to human community experimentation would arise with respect to private persons. The two principal issues involved revolve around the notions of consent and tort liability for harm. We must also consider the possible issue of criminal liability. To get a clear perspective we can

examine the medical research findings that apply to
our concern. There are certain "fixed" notions:

(a) There really is no body of "law" dealing with
    human experimentation.
(b) "In the treatment of the patient there must be no
    experimentation."[11]
(c) Where experimentation takes place it must be
    based on the knowledge and consent of the
    person affected.

These three generalizations imply the most pertinent
legal and ethical issues related to research with
human subjects. Items a and b are seemingly in
direct conflict, but actually this is not so. Scholars
have searched for guidance concerning the legiti-
macy of experimentation with humans, but have
found no existing law dealing with the subject.
There are not even any standards that define the
acceptable parameters of this sort of research.[12]

In large measure this situation exists because
there has not really been any planned scientific research in this area to merit litigation. When issues
have arisen in the past they have been concerned
primarily with claimed injuries resulting from the
methods of treatment employed.

No reported court decision has considered research
specifically in terms of the right and liability of a
trained professional to use a living patient or normal
subject as a means of discovering new knowledge not
necessarily of direct benefit to that patient or subject.
None of the cases that have actually come before the
appellate courts have involved a real scientist observing
the proper precautions and giving primary consideration to the welfare of his patient.[13]

While of course there could be legitimate dispute
among scientists about the proper techniques and

controls to be used in an extraterrestrial experiment, it seems that under carefully defined circumstances, research might be legally feasible. In order to insure against liability on the part of the sponsoring party or agency, any high-risk experiment would have to be sanctioned within the guidelines of our existing legal framework. Dietrich states that "legally, it is difficult to regard psychological experimentation as other than an intentional invasion of the subject's interest in peace of mind, an interest entitled to independent legal protection in the absence of a privilege conferred by society and the subject's consent."[14]

Before examining the ethical issues involved in experimentation with the knowledge and consent of the person involved, the historic precedents relating to nonexperimentation should be explained.

The rule that "in the treatment of the patient there must be no experimentation"[15] relates precisely to possible liability, but does not in itself preclude *some narrow limit*. This narrow limit is based upon scrupulous and clear procedures and carefully drawn concepts of consent. Without going into the details of the antecedent cases, their main principles revolve around the use of tried and proven methods of treatment. Indeed, a recent landmark case decision on psychosurgery was predicated on the issue of an unaccepted and unproven procedure. Deviation from such "approved" methods becomes a matter of personal risk for the practitioner (*Slater v. Baker*, 2 Wils. K.B. 359, 95 Eng. Rep. 860, 1967; *Carpenter v. Black*, 60 Barb. 488, N.Y. Sup. Ct., 1871). Although this has been the prevailing view and the burden has been on the experimenter to justify his methods and procedures, there is another dimension to the issue. One significant state case recognized that if progress were to take place, some experimentation would have to be permitted. The court in this instance felt that consent and knowledge were prerequisites and that the experiment should be as closely related as possible to traditional methods (*Fortner v. Koch*, 272 Mich. 273, 261 N.W. 762, 1935).

In addition to the ethical concerns of knowledge and consent, which are highly related to legal thinking, there are two other pertinent legal issues. The first is experimentation where it is possible to show, as is the case of "experiments" with kidney transplants, that for one of the parties involved, there is a lack of benefit, perhaps even a positive harm. It is conceivable that the psychological aspects of these experiments, properly assessed, can mitigate liability.[16]

The second related issue is the development of a concept called "liability without fault,"[17] that is, society would recognize the necessity of types of research that would traditionally be illegal, and instead of preventing the research or placing the researcher in a hopeless personal situation, the government would take the physical and monetary responsibility for harm as a matter of social responsibility. Of course, careful controls and consent of parties to the experiment would be essential prerequisites.

Throughout this examination of the issues related to human community experimentation, the concept of consent has been raised at every turn. It is probably the focus of the ethical concerns with extraterrestrial-type experiments. The extensive discussion of the ethical aspects of human experimentation began primarily as a result of the Nuremberg trials. In those trials for the first time in history war crimes were made the basis of a postwar trial of persons who committed "crimes against humanity." Human experimentation of the most vile sort was included in these crimes. The experience with the Nuremberg trials resulted in formulations by professional groups of ethical codes all over the world. In addition to these formulations of principles, religious and philosophical leaders and empirical laboratory scientists have attempted to deal with the vast series of ethical questions posed by human experimentation.

With respect to knowledge and consent, it clearly is the responsibility of the investigator (qua anthropologist, political scientist, psychologist, etc.) to explain to a potential community participant all of the possible consequences of the experiment. In addition to full disclosure prior to the onset of an experiment there should be agreement concerning the conditions under which an experiment (community) could be halted by the participant. Inherent in the community, there would have to be a reserved freedom for the individual to change his mind. ("Stop the extraterrestrial community—I want to get off.")

An experiment with an extraterrestrial balanced enclosed life space community raises some other important issues. Does an individual have a right to consent for the potential new generation that might be born within the vivarium community? Is it really possible to explain eventualities well enough so that persons can truly know what it is that they are consenting to? We suggest that there are limits to a person's right to consent and that our traditional legal concepts may be inadequate for the era of the extraterrestrial community vivarium.

As behavioral scientists whose chief concern lies within the realm of culture and the role of law in social processes, we would like to conclude that there is a vast difference between cultural norms, legality and cultural desirability. We often have a tendency to look at normative culture and codified law in a rather mechanistic fashion. The fact that culture and the law per se may not stand in the way of certain types of experiments or adventures with humans does not mean that they are truly essential or not without other and perhaps more important implications.

With respect to extraterrestrial communities, the most important question is the consideration of possible methods of acquiring the data without recourse to such a drastic and costly experiment.

The total environment and commitment of the community would place incalculable strains on a person's freedom to change his mind during the course of an orbit. Thus the balanced life system that is to be investigated or tested seems to be inconsistent with what to us have been important basic values. As a people we have believed in enhancing the dignity of the individual, maximizing his freedom, using the state as a service agent rather than as a controlling master. In addition to placing severe strains on these values for persons entering into the community, long-term existence in such a community would pose an important threat to basic rights, both procedural and substantive, since protections would not be afforded to future generations.

Science is a tool of humanity. It can be used to aid in the understanding, control and manipulation of the environment. The basic values to be used or applied with reference to any scientific enterprise are not intrinsic to the scientific activity itself, but must derive from some other source. In our cultural-governmental system we have attempted to minimize human conflict through the utilization of public, politically determined values. In our view, the extra-terrestrial community would make such application impossible and would obliterate whatever line of demarcation remains between what is public and what is private.

Notes

1. A symposium was held in 1965 at the American Psychological Association meeting on the theme of "Balanced Life Systems: A Test Tube for Life Science Research." The organizer was psychologist Louis Aarons. In 1966 V. V. Parin, a Soviet psychologist, was the organizer for a Symposium on "Psychological Problems of Man in Space." This was held in connection with the XVIII International Congress of Psychology, Moscow, U.S.S.R. To the best knowledge of the authors, none of these papers has been published.

2. Lewis Mumford, "Utopia, the City and the Machine," *Daedalus,* Spring 1965, pp. 271–292.

3. S. F. Nadel, *The Foundations of Social Anthropology* (Glencoe, Ill.: The Free Press, 1951), p. 222, quoting T. Parsons, *The Structure of Social Action* (1937), p. 743.

4. Oscar Lewis, "Controls and Experiments in Field Work," in *Anthropology Today,* ed. A. L. Kroeber (Chicago: University of Chicago Press, 1953), p. 463.

5. D. G. Haring, "Anthropology: One Point of View," in *Personal Character and Cultural Milieu,* ed. D. G. Haring (Syracuse, N.Y.: Syracuse University Press, 1956), p. 11.

6. D. G. Haring, "The Scientific Study of Social Phenomena," in ed. D. G. Haring, op. cit., p. 108.

7. Ibid., p. 115.

8. Ibid., p. 108.

9. Forde, C. Daryll, *Habitat, Economy and Society* (London: Methuen and Company, 1949), pp. 463–465.

10. W. H. Johnson, "Civil Rights of Military Personnel Regarding Medical Care and Experimentation Procedures," *Science* 117 (1953) : 212–215.
11. E. L. Cady, "Medical Malpractice: What About Experimentation?" *Ann. West. Med. & Surg.* 6 (1952) : 164–170.
12. I. Ladimer, "Ethical and Legal Aspects of Medical Research on Human Beings," *J. Pub. Law* 3 (1954) : 467–511.
13. S. M. Sessoms, "What Hospitals Should Know About Investigational Drugs: Guiding Principles in Medical Research Involving Humans," *Hospitals* 32 (January 1, 1958) : 44ff.
14. D. P. Dietrich, "Legal Implications of Psychological Research with Human Subjects," *Duke Law Journal,* Spring 1960, pp. 265–274.
15. Cady, op. cit.
16. J. W. Curran, "A Problem of Consent: Kidney Transplantation in Minors," *New York University Law Review* 34 (1959) : 891–898.
17. Dietrich, op. cit.

Further Readings

Beecher, H. K., "Some Fallacies and Errors in the Application of the Principle of Consent in Human Experimentation," *Clin. Pharm. & Therap.,* 1962, *3,* 141–146.

Cahn, Edmond, "The Lawyer as Scientist and Scoundrel: Reflections on Francis Bacon's Quadricentennial," *NYU Law Review,* 1961, *36,* 1–12.

Esecover, H.; Malitz, S.; Wilkins, B.; "Clinical Profiles of Paid Normal Subjects Volunteering for Hallucinogen Drug Studies," *Am. J. Psychiat.,* 1961, *117,* 910–915.

Hatry, P., "The Physician's Legal Responsibility in Clinical Testing of New Drugs," *Clin. Pharm. & Therap.,* 1963, *4,* 4–9.

Hill, A. B., "Medical Ethics and Controlled Trials," *British Medical Journal,* 1963, *1,* 1043–1049.

Kevorkian, J., "Capital Punishment of Capital Gain," *J. Crim L. Crimin. P.S.,* 1959, *50,* 50–51.

Kidd, A. M., "Limits of the Right of a Person to Consent to Experimentation on Himself," *Science,* 1953, *117,* 211–212.

Perlin, S.; Pollin, W.; Butler, R. N.; "The Experimental Subject: The Psychiatric Evaluation and Selection of a Volunteer Population," *A.M.A. Arch. Neurol. & Psychiat.,* 1958, *80,* 65–70.

Richmond, J. B., "Patient Reaction to the Teaching and Research Situation," *J. Med. Education,* 1961, *36,* 347–352.

Shartel, B. W., and Plant, M. L., *The Law of Medical Practice* (Springfield: Charles C Thomas, 1959) .

Smith, E. E., "Obtaining Subjects for Research," *American Psychologist,* 1962, *17,* 577–578.

Swift, J., *Gulliver's Travels* (London: Oxford University Press, 1954) .

5.
Terra-Lune:
A Frontier
City-State

BILL GERKEN, JR.

In the eyes of some, there was scant reason for cele-
bration on both the last day of the second millennium and the first day of the third. Nevertheless,
there was an official ceremony at Terra-Lune to mark
the passing of the old and the dawning of the new.
The purpose was to formally adopt "Terra-Lune" as
the new name of the lunar city-state. Although the
name has been in conversational use for more than a
year, the original name, Grissom-Komarov—usually
shortened to "G-K"—had been retained through the
year two thousand. At the same time, the lunar city
ceased reckoning time in accordance with earth
calendars. 2001 A.D., then, has become the year 1 T-L.

As difficult as it was to sever the old ties to the
mother planet, everyone knows that no one can re-

turn to earth for an extended stay for a century or
more. Actually, there are few who have either the
time or the inclination to mourn the earth. More
than half the lunar population has been here for
more than ten years. Long ago they accepted the
moon as their natural home. There are, of course,
those who arrived here during the hectic last days of
earth's civilization. They carry with them the still
fresh memories of earth, some of which contrast
sharply with the new realities to be faced on the
moon. They regard the moon as a temporary home, a
"bomb shelter" of sorts, to be used only as long as
necessary and then discarded, when civilized life can
safely be reestablished on earth. The "Returners,"
as they call themselves, are few in number, and there
seems to be little chance that they will be able to
sway the opinion of a majority of the lunar
population.

Grissom-Komarov—named for two spacemen who
lost their lives in the early space program—was established in 1984 by the Lunar Alliance, an organization
comprised of the spacefaring nations (the United
States, Russia, Japan, the United Kingdom, Canada,
France, West Germany, Italy and the People's Republic of China). In its statement of purpose, the
alliance set forth this goal: "To explore the moon
for minerals needed to sustain life on the earth; to
extract, process and conduct research on these materials so that they may be used in the most efficient
and economical manner; and to manufacture on the
moon and in space those items for which the local
environments are most suited."

As with most official documents, what was left
unsaid was far more important than the words that
appeared in print. The documents implicitly pro-

vided for the establishment of a permanent base on
the moon. This base—Grissom-Komarov—was, in
the most secret papers covering the intergovernment
accord, to become a city-state, independent of earth.
It was hoped that the fledgling city could become self-
sufficient quickly enough to give humanity a second
start when the apparently inevitable collapse of
earthbound civilization came.

The first people to sever their ties with earth
were, of necessity, many of the men and women who
had flown and worked in space during the preceding
eleven years. As the core of the lunar surface team
they built the first shelters and installed life-support
systems. They carefully adapted the best equipment
earth could provide to the requirements of the demanding lunar environment. Their families, although
left behind, were not forgotten. Those lunar workers
whose contracts came up for renewal were free to
take their chances and return to earth. Those who
opted to have their families join them on the moon
instead entered into agreements with their spouses
and the Lunar Alliance, whereby the alliance would
educate the earthside family members in fields that
would make them useful and productive members of
the lunar staff. Since the final goal was a complete
society, great diversity was allowed in the selection of
studies, with one stipulation: those who chose fields
deemed to be of indirect and long-term value would
not go to the moon as long as all the available
passenger space could be filled with people possessing
immediately useful skills.

As with any human arrangement, the conditions
were not agreeable to all. There were a few resigna-
tions, several divorces and a surprising number of
hasty marriages. The marriages joined men and
women in the lunar training program who hoped to

leave earth together, thereby avoiding the emotional stresses of separation for what might be a year or more.

The first permanent residents of the moon, then, were a mixture of scientists, pilot astronauts, engineers and technicians. Each had necessary second skills—in medical areas, for example—to see them through the first year. They were a highly skilled, highly motivated group, used to challenging the unknown and accomplishing what others said was impossible. They worked well together, despite occasional personality clashes and the conditions under which they lived and worked.

Safely out of reach of the multigovernment bureaucracy which had spawned it, the initial lunar team quickly developed a loosely structured organization ideally suited to their tasks and temperaments. On-the-job cross-training went far beyond their prelaunch schooling. Combined with their drive to succeed and the willingness to help whenever and wherever needed, their broad-ranging abilities carried them through the early stages.

Forty-one men and sixteen women made up the first lunar team, the imbalance between the sexes being primarily due to the smaller pool of available spaceflight-qualified women at that time. Although only two couples were married, the relaxed sexual standards of the eighties made the isolation from earth more bearable for all. Then, too, since Grissom-Komarov's ultimate end was to be the home of a diverse and rounded population—men, women and children, families and singles, young and old—the presence of women from the first days proved invaluable in keeping the city from assuming an austere, militaristic, all-male atmosphere.

During the first seven years, while the Lunar

Alliance's avowed purpose of exploration, mining, processing and manufacturing for earth's use was still widely accepted, the lunar workers spent an average of one-year contract shifts on the moon, followed by a three-to-nine-month rest and training cycle on earth. The variation in the period spent on earth was deliberately made flexible so that workers returning from G-K could reassess their feelings toward remaining on earth permanently or convincing their families to work toward joining them on the moon.

When the first replacements arrived, nine months after the initial landing, the newcomers immediately began to tie in with the various project leaders and the base's two administrators so they would be fully capable of carrying on once the key members of the first team had all left for earth. The last members of the initial lunar surface team remained at G-K an additional three months to insure the smoothness of the transition between the old and new work forces.

With the first shelters in place, the life-support systems functioning, the first crops planted, grown and harvested, and primary on-site surveying completed, life at G-K was far more orderly for the recently arrived work force than it had been for their predecessors. The second year saw the city double in population since, for the first time, people without prior spaceflight experience or lengthy training journeyed across the void between planet and satellite.

The next five years saw Grissom-Komarov blossom from a crude frontier outpost to a dynamic town, cosmopolitan in character thanks to its multinational citizenry, but unlike any of earth's great cities. The heavy emphasis on science and technology gave G-K more the flavor of a university town or a think tank than of a city, yet here too there was a difference. Because the city's principal product was not students or reports or knowledge, but Grissom-Komarov itself, there was an activity and sense of purpose often lacking in the university environment. Knowing that humanity's survival could well depend on their success or failure united efforts in a way unmatched in earlier international ventures.

The lunar city was dependent on its Terran sponsors for nearly everything during those first seven years. Although inroads were being made in many areas, important basic materials were still shuttled by rocket from earth. In 1990 G-K for the first time was able to feed its people solely on food grown on the moon. Even then, a few luxury items were imported from earth for celebrations and festive occasions.

Grissom-Komarov became self-sustaining first in the production of its own atmosphere and water, which were extracted from the lunar soil and rocks. Since both systems were closed cycles, relatively little additional air and water were needed to make up for leakage to the vacuum outside the city's domes and plumbing.

By 1990, the population had grown to slightly more than a thousand, with women making up approximately one-third of the total. During the preceding four years, the annual influx had remained constant at about two hundred people a year. The Lunar Alliance wanted G-K to become self-supporting as soon as possible, in view of the deteriorating conditions on earth. To achieve that end, the earthside organization limited the number of people added to the lunar work force each year, according to a continuously updated computer program which projected probable self-sufficiency dates for the lunar city against population figures. As work progressed on the moon, and as more people arrived from earth, new projections replaced older ones. The 1990 year-end projections showed that if no additional people joined those already at G-K, the city should achieve independence by 1995. If the lunar population continued to grow at the past rate, it would not become self-sufficient until well after the turn of the century.

In early January 1991 the Lunar Alliance met to make one of its most important decisions. The population of Grissom-Komarov would not be further increased until the city became self-supporting, except in the event of special need, or to reunite husbands and wives under the terms of existing agreements. Those already on the moon were in full accord with the decision, and had been among the earliest to push for its adoption.

The Lunar Alliance itself was going through an extremely critical period at that time. Word had finally gotten out that the city on the moon was just that; it was not a remote resource-gathering station for the nations of earth. The fact that G-K did indeed send every shuttle back to earth carrying automatically mined lunar minerals was instantly forgotten in the outcries of treachery and deceit from the press of every nation. Amazingly enough, the governments involved stood squarely behind the only truly functional international body left in exis-tence. Leaders paused in their domestic bickerings and in their hot and cold wars to support the Lunar Alliance. The controversy died slowly, and by mid-year it had been replaced by a dozen new scandals, wars and crises, all much closer to home than the moon.

There can be no doubt that the survival of Grissom-Komarov during that period was due to the spirit and untiring work of the earthside Lunar Alliance staff. Knowing full well that their own skills were not those that would guarantee them lunar citizenship, they continued to carry the burden of selecting from earth's four billion people the group that would be allowed to start anew on the moon. Their tremendous drive and dedication, right through to the last shuttle flight, will stand forever as the highest personal sacrifice in the annals of human history.

Grissom-Komarov's character did not change appreciably during the last half of the eighties. The population remained heavily scientific and technological, although there was a growing number of people engaged in manufacturing, teaching, agriculture and various services. The rapidly expanding population required more formal organization than had been needed in the first years, when everyone concerned with a problem could gather in one shelter to discuss it. Designed into the city's basic systems was an extensive two-way communications network of the kind proposed for American cities just prior to the mid-seventies energy crisis. A combination of cable television, telephone and central computer, the network was as ideally suited to participatory government as were the citizens of G-K. Well-educated and intensely interested in anything which might have a bearing on their situation, the lunar city dwellers reveled in the opportunity to take an active part in city affairs through the medium of instantaneous electronic polling.

Although all could have a say in government, the citizens of the moon realized that a few people ought to be ultimately responsible for the city's management. The lunar environment was still alien, harsh and unforgiving, and with conditions on earth worsening with each newscast, they knew that they must take every precaution to prevent the simple oversight that could cause their entire endeavor to fail.

The system of government was designed by the Lunar Alliance's social scientists to fit the conditions and personalities at Grissom-Komarov. With only minor changes, it is still in use today. Overall administrative responsibility belongs to the city coordinator. A background in systems and synergistic work is the prime requisite for the city coordinator, who is chosen by the systems coordinators. The city coordinator selects his or her own deputy. Both serve for six years, or longer at the discretion of the systems coordinators, in additional two-year increments. Only the city's management board, by a three-quarters vote, can remove a city coordinator from his position.

Terra-Lune's management board is comprised of seven systems coordinators and seven citizens-at-large. Systems coordinators, one from each of the city's functional organizations, are the administrative heads of their respective systems. At present, the systems include: life support, manufacturing and distribution, research, development and engineering, humanities, education, and health. The humanities system encompasses the arts and the social sciences, while the life-support system spans agriculture, power, atmosphere, water and waste materials.

The seven board members known as citizens-at-large are elected by a vote of the total adult population. To insure diversity of viewpoint on the board, there is a provision that no more than two citizens-at-large may come from any one system organization.

The board members also act as ombudsmen for their fellow citizens.

City government was established in 1988, when Grissom-Komarov's population reached six hundred. Until then, the city coordinator had been the city's only official, working with ad hoc committees as needed. With the total number of lunar citizens still well below two thousand, no serious political factions other than the Returners have formed, and no change in the system of government is anticipated for at least thirteen years. Terra-Lune's Constitution calls for a review of the governmental system every twenty-five years, to ensure that government remains relevant to the city's needs. Such reviews can take place at more frequent intervals, by two-thirds vote of the management board or of the adult population.

At the end of 1996, Grissom-Komarov's citizens numbered 1,286. The city had become self-sustaining in late 1994, beating the computer predictions through a massive effort by the lunar residents. Knowing what loomed ahead for the people of earth, those on the moon continued to restrict their usage of all materials, stockpiling them as security for the future.

Once self-sufficient, the roles in the Lunar Alliance reversed. The earthside staff became dependent on Grissom-Komarov for their direction. A special Earth Planning Group was formed on the moon to guide preparations and work during the hectic last three years. First, the children of those at G-K were brought over from earth. The lunar citizens did not fully anticipate the despair of those on earth. Some youngsters were kidnapped and others substituted in their places. Some were spirited away by the relatives on earth with whom they had been staying, who feared losing the young ones in what they termed, "this mad folly."

A series of top-secret messages from Grissom-Komarov to the earthside staff during the fall of 1996 laid the foundation for the final flights to the moon in the first months of 1997. Slowly, the most dedicated and energetic of the earthbound Lunar Alliance workers were transferred to the launch centers, along with their immediate families. Between Christmas and New Year's Day, each was asked if they would accept lunar citizenship. Those who refused were returned to their homes under close security surveillance. The others rode earth's final nonmilitary rockets into space in January and February of 1997.

The offer had not been solely a matter of gratitude or charity on the part of the lunar city. The people and their talents had been carefully screened. Within limits, the greater the moon's population, the better the city's odds for survival. Extra supplies had been set aside for a number of predicted late arrivals, in order not to strain the lunar economy beyond its ability to recover.

In spite of the elaborate secrecy surrounding the preparations, there were some problems. When contact was finally lost with earth, 215 of Grissom-Komarov's 1,757 residents were uninvited guests. This posed a problem for the tightly defined lunar economy. They were welfare cases in a community without welfare agencies.

G-K's highly educated, multinational citizens were continually confronted with the realization that the settlement was in every sense a frontier city, now cut off more completely from its past than any other in human experience. Harsh realism prevailed.

For some time, the management board had been considering renovating one or more of the outlying resource survey base camps that had been used during the first years of lunar mining operations. The main reason for delaying the start of work at the camps had been the lack of personnel for the job. The unwanted immigrants presented a possible work force for the task. They were transported to the camp's two crude buildings where they were to erect permanent domes, systems and a radio relay tower on nearby mountains with the equipment with which they were provided.

The outcasts complained that the city's actions were tantamount to murder. The city replied, "All we ask is that you convince us that you deserve to be among us. You have more than enough supplies and materials to last, and if you want to survive, you can. When the tower is complete and you contact us through it, you will be welcomed into our society. The choice is yours."

Three weeks later, Grissom-Komarov's radio towers picked up a message from the survey camp. Some of the survivors asked that they be brought back to the city. Others requested that they be loaned additional supplies until they saw whether or not they could survive on their own. Their request was approved by the management board with the stipulation that they limit themselves to food production and small electronics assembly, the city's two most critical needs at the time. A small team of agricultural and electronics experts moved out to the camp with the necessary equipment. Of the hundred and seventy-two men and women who had been taken to the camp originally, eighty-five opted for life in the city and fifty-one stayed at the camp. Little was said of the thirty-six who died during those three weeks, other than that they did not carry their share of the burden. The deaths, far closer to home than earth's billions, were a sobering reminder to all that our new home's demands were not to be ignored without incurring exacting penalties.

Even before immigration from earth had been completed, the city's population had become more diverse than it had been in the eighties. G-K's citizens now included a tiny number of professional artists, writers, performers and theologians. Much of earth's cultural heritage had been preserved on microfiche by the Lunar Alliance staff, and delivered to G-K on regular shuttle flights. The last years of earth's modern civilization inspired G-K's artistically talented residents. Now, for the first time, the dry technical reports, photographs and films came alive as paintings, sculpture, poems, stories and plays. The old national origins and customs were not forgotten, but the tragic events of the second half of the century cast them in a new light. Once-fierce prides were muted to the warm, comfortable feelings one has for favorite friends and relatives. The Christmas and Hanukkah holidays were combined into a Season of Giving, lasting for two weeks in December.

Although Terra-Lune's theologians still conduct services regularly for those who find them comforting, attendance has remained small. To most of the people here, religion is a highly personal matter, not requiring formalized organizations and meetings. Even at the services, the theologians, by common agreement, have been personalizing their respective creeds, making them more relevant to the daily lives of the faithful.

The Jesuit mathematician who serves the city's Catholic and Russian Orthodox citizens went through a trying period in late 1997, when he realized that he could well be considered both Pope and Patriarch of the Catholic Church. For a week, the only thing that could bring a smile to his lips was the knowledge that at last the church had been reunited. He finally decided to settle the matter by announcing, after consultations with his peers and his congregation, that there would no longer be an official head of the church.

By the end of the year 2000, Terra-Lune had grown from a collection of temporary shelters to a complex of large, permanent geodesic domes, connected by long, tunnellike corridors. Each dome has its own atmospheric regeneration system, and each system has twice the capacity needed for a single dome. In the event of failure in any system or dome, the load can temporarily be carried by the surrounding domes until repairs can be completed. VEN-Suits (*V*acuum *EN*vironment, or pressure, suits) are no longer stored in residences, just as parachutes were not carried on commercial airliners on earth. None of the catastrophic failures predicted in the city's early days have occurred, and the citizenry is free from the paranoiac feeling that nature is trying to destroy this human outpost.

Although some of the domes are devoted completely to a single activity, like farming, most are designed according to one basic pattern. At the outermost edge, where the roof curves down to meet the ground, is a ring of residences. Apartments vary in size from one to three bedrooms, depending upon the number of people living together. Bedrooms are compactly designed, much on the order of the railroad and ocean-liner sleeping accommodations of fifty years before. When not in use, beds fold away, turning the room into a personal sitting, work or study area.

A family room and bath complete most of the apartments. A few of the residences in the most recently completed dome also have an extra room which can double as an office or a dining area. Even the three-bedroom apartments have only one bath, and there are no luxury apartments for the city or the systems coordinators. Terra-Lune's society is still sufficiently classless not to require such status symbols, although the feeling is growing that, by the next generation, some form of reward for extra service may be necessary. For the present, the original dedication and pioneer spirit are more than adequate to support those who take major roles in lunar government.

Since the first days, all food has been prepared in central kitchens in each dome and eaten in cafeterias. Reductions in energy usage, apartment size, and time saved in communal food preparation dictated this approach. To compensate for the loss of the traditional custom of home cooking and dining, two new arrangements were introduced into life in Terra-Lune. The cafeterias were designed as series of rooms of varying sizes, to personalize them as much as possible and remove the feeling of eating en masse. The smallest rooms have folding screens which allow individual families to dine alone, provided they do not abuse the privilege through overuse. For couples and those who prefer to eat alone, the larger dining rooms have booths along the walls.

Realizing that communal food preparation would waste or stifle much of the city's culinary talent, the kitchens have been thrown open to anyone interested in cooking. Working together with the professional kitchen staff, men and women can prepare their favorite dishes for their families and friends, while the staff expands the recipes to include them on the daily menu of that cafeteria.

Proceeding toward the center of a large dome from the residences at its perimeter, one crosses through a wide band of growing plants. While most of Terra-Lune's agricultural crops are grown in the farming domes under closely controlled conditions, many fruits are grown in these park rings in each of the main domes. Ripe oranges, apples, peaches and other fruits can be picked from the trees and eaten as one makes his or her way across the city. The park rings also add a touch of earth to the austerity of the moon, separating as they do the apartments from the other areas of activity.

Clustered at the center of each dome are offices, shops, laboratories, support systems buildings, manufacturing, storage and assembly areas, infirmaries, educational and recreation complexes. The two newest domes have an underground level which contains their support systems, warehouses, manufacturing and assembly facilities. No buildings rise more than five stories above the ground, even though the domes are high enough near the center to accommodate buildings with two more floors. The open space above the buildings gives Terra-Lune the feeling of spaciousness it otherwise lacks.

There are no elevators in the city other than those used to move heavy equipment and loads of supplies. All multistory buildings have "bounce shafts" instead of stairs for travel between floors. These elevators take advantage of the low lunar gravity by permitting one to float gently downward, stopping at the desired level by grasping a handhold. To go up, one either springs off the small platform at the floor he is on, or bounces upward off the trampolinelike webbing placed just below the ground floor. Handrails along the length of the shaft help one stop at the proper floor. Until the sports dome was built last year, the city's children used the tallest buildings' bounce shafts to perform low-gravity gymnastics.*

Terra-Lune's controlled climate allows many activities to be conducted "outdoors" (under the dome, but not inside a building) that were necessarily confined to indoors on earth. In almost every field of endeavor, some people choose to work beyond the walls of interior buildings. It is not unusual to find desks, noncritical assembly areas, shop stalls, hospital beds and food-preparation tables scattered among the trees and bushes. One cannot cross a dome without seeing school discussion groups, engineers and shopkeepers going about their daily work in the open air.

For the most part, dress is casual in Terra-Lune. Controlled variations in air temperature are such that one will usually choose lightweight clothing. The first years of Grissom-Komarov saw all clothing issued in semimilitary styles. Modifications began to appear according to personal taste and national background. By 1989 many people were designing and making their own clothes. The following year, the management board approved plans to diversify the kinds of clothing available, and to offer them in the city's shops. This has probably been the most popular decision in the board's history. Now many people own a set of clothes reflecting their national

* The author is indebted to Dr. Isaac Asimov for the concepts of both the "bounce shaft" and the lunar gravity sports and recreation dome. Dr. Asimov earlier described similar facilities in his novel *The Gods Themselves.*

heritage, which they wear on holidays and other special occasions. Terra-Lune is still far from being a society where fashion dictates what people will wear, but the drabness of the past is gone.

All work in Terra-Lune is divided among the seven systems represented on the management board. Two systems, life support and manufacturing and distribution, employ more than 50 percent of the working population, since their efforts directly affect the daily lives of everyone in the city.

The working population includes every person over the age of seven. Children are given tasks equal to their skills. They operate the park ring watering systems, sort materials for recycling and help with harvests. They start at only an hour a day, with the time and task complexity gradually increasing until they reach adult status, at age fifteen. Jobs are rotated frequently to enlarge experience and to prevent boredom. This forms a valuable part of the youngsters' education, giving them an early introduction to the adult world, and at the same time the sense of usefulness which had for years been lacking in countries like the United States. It also provides them with concrete examples of the importance of cooperation in the complex, interdependent society in which they live. Finally, by putting children in daily contact with adults at work, their eventual career choices become easier to make.

Beginning at age sixty, one has the option of selecting fewer work hours along a scale which decreases to an hour a day at sixty-seven. This "retirement plan" can be delayed as long as one likes, health permitting. Senior residents continue to work at least an hour a day as long as they are able. This policy is necessary because Terra-Lune's existence is not yet firm enough for the old and the young to be separated from the mainstream of community life. Many in the humanities system hold that a major factor in the fall of Terran culture was such a separation, and they have been strongly urging that we do not make the same mistake on the moon.

In Terra-Lune, we have reached the point that earth's developed nations were striving for during their last decades: a moneyless society. Resources are allocated according to need when they exist in sufficient quantity. When there is a scarcity, a system of voluntary rationing is adopted. The city has not yet been faced with an unwillingness to work on the part of its people, and the example of the unwanted immigrants of 1997 remains as warning to any who might consider malingering.

Agriculture is the most important function of the life-support system. Over 75 percent of Terra-Lune's food is grown in single-purpose agridomes, equally divided between hydroponic and lunar soil culture. Each agridome is temperature-, humidity- and light-controlled to provide optimum growing conditions. Computer sensors attached to a sample population of plants respond to their needs by automatically activating the water and nutrient systems.

Both the atmospheric and water subsystems of the city are closed cycles. Like the agridomes, each city dome is individually controlled for temperature, humidity and light. The slight variations between domes and from one day to the next relieve the constant sameness of a closed environment. The interdome corridors act as buffers between climates, and it is not unusual to hear residents of different domes exchange comments about the "weather."

All water is purified for reuse. Terra-Lune's water is so clear and pure that the initial dislike of drinking water which once contained urine no longer exists.

Solar and nuclear energy provide all the city's power. During the two weeks of sunlight each month, the sun's energy is collected by huge solar panels stretching for miles across the lurain. These solar-array "farms" stockpile energy not immediately needed in giant storage batteries for use during the two-week-long lunar night. There are still not enough solar arrays and storage cells to fully provide for all the city's requirements throughout the half-month of darkness, as the demand necessarily increases to counter the moon's cryogenic nighttime temperatures outside the domes. To make up the difference, a nuclear power plant located several miles from Terra-Lune is used. This plant dates back to the 1980's, when it supplied virtually all the city's power, and is expected to remain in service for the foreseeable future.

Sewage, once the water has been removed, is treated to remove bacteria and then broken down into usable fertilizers for the crop subsystems. What remains is dried and compressed for use as a fuel in those manufacturing processes which still require open-flame sources. Other waste materials are segregated at the point of discard into basic groups (metals, plastics, cloth, paper). From these points, they are taken to the appropriate reclamation plant.

Food wastes are processed to feed Terra-Lune's small but well-loved animal population. Several kinds of dogs, cats and birds can be found in the city: another effort to retain some of earth's flavor. All the dogs and cats are owned by individuals, while the birds are considered a community asset. There is always a waiting list for puppies and kittens, due in part to the necessary limitations placed on breeding. Any excess food waste is put into the sewage subsystem for reconversion.

Equally important to Terra-Lune's daily existence is the work of the manufacturing and distribution system. Somewhat fewer people work in M and D, due to the automatic machinery used in most of the assembly facilities. The greatest burden for the city's survival falls on M and D, for its task is defined as "The ability to produce everything, other than what is supplied by life support, necessary to the continuation of a modern, civilized life style." Many of earth's luxuries, such as alcohol, tobacco, cosmetics, facial tissue, closets full of clothes, are absent from Terra-Lune. There is, however, a reasonable degree of comfort for a pioneer town in the middle of a vast wilderness.

The workers in "manufacturing" produce a constant flow of garments, furniture, construction materials, replacement components, modules of every size, shape and description, control knobs, computer assemblies, rocket engines and razors—the list is endless. There are also skilled craftspeople who repair worn items, making their own parts when a stock has not yet been created. These same people disassemble and rebuild the city's critical medical equipment, the one-of-a-kind units which must be kept functioning until duplicates can be built. They are aided in these tasks by a microfiche file of the blueprints used to manufacture everything ever shuttled from earth to Grissom-Komarov.

The distribution subsystem handles delivery of goods to the proper locations throughout the city (excluding food items, which stay within the life-support system from farm through cafeteria to sewage). Items produced for private use, like clothing, are transported to the many shops around the city, so that the residents may make their own selections. Admittedly not as efficient or economical as a direct distribution to individuals, the shop system was another deliberate compromise with the frontier to increase the personal freedom of the citizens. Such compromises make life's other deprivations easier to bear. Quiet electric carts are used to move M and D's produced goods from assembly facilities to the shops and storage areas.

Education is a lifelong process at Terra-Lune. It begins informally in the child-care centers, as soon as toddlers are able to recognize the difference between colors, shapes and sounds. Indeed, with the open-air classrooms and the daily work break for the older children, informality is the cornerstone of learning on the moon. This is helped immeasurably by the high average education level of the lunar citizenry, the relatively small number of children to be educated and the availability of a central computer system. One of a child's earliest toys is a pocket-sized computer keyboard which can be plugged into the central educational computer. At first, the youngsters find sufficient delight in pressing the number/letter keys and watching the display lights wink on and off. Shortly, they are shown how to tie in to the main computer and how to play simple games with it. By the time they begin to receive instruction from educators, the computer is their constant companion. Instead of memorizing times tables, they learn what the multiplication operation is, and how it should be used to solve problems.

As on any frontier, practical learning takes precedence over the intellectual or the aesthetic. The daily work break and visits to every other area of activity in the city play a big part in the youngsters' preparation for the responsibilities of adulthood. In spite of these considerations, the children of Terra-Lune receive a complete and well-rounded education. Here, too, the adults, with their wide range of talents and interests, are a major asset. Work schedules are flexible enough to permit adults to share with youngsters their knowledge of the arts and sciences.

There is no discernible end-point to education in Terra-Lune. No formal degrees are granted, and anyone's level of achievement can be quickly learned from the central computer memory banks, provided only that a need for this information exists. It is not usual for people to change career fields, but the opportunity to do so is available to anyone desiring a change. It is felt that the early introduction of children to work, and its constant closeness to them, is responsible for helping them make more valid career choices than was possible on earth.

Terra-Lune's health system emphasizes health protection as well as illness prevention. The underlying theory is that it is both easier and more productive to keep the citizenry in good health than merely to react to disease and illness. Health care is active instead of passive, as it was during most of human history. Periodic examinations, a balanced diet and daily exercise all play a part in keeping the residents healthy. Because the city is a closed community, most contagious diseases have been eliminated. People are not in perfect health, but the number of illnesses that affect the human body are slowly decreasing. It is the limited number of medical researchers and the insufficient amount of equipment that delays any real advances. The trauma of earth's dissolution and the isolation of Terra-Lune have been responsible for a number of cases of mental illness, but modern treatment and new drugs have helped them considerably.

In the year 2000, the city marked the birth of the three-hundredth lunar-born baby. As with most other things on the moon, population growth follows a plan. Women wishing to bear children are asked to advise the health system office, which keeps track of the number of current pregnancies. The women are then told either to go ahead as they wish, or to wait a few months, depending on how busy the maternity infirmary is scheduled to be at the time of birth.

A principal concern of the humanities system is the study of the city itself. Here, theory and experience meet, for the other systems are constantly requesting suggestions and advice as to how they might improve their particular services to the city and the citizenry. In this respect, Terra-Lune is a social scientist's dream-come-true, for he or she has an entire community with which to work. Occasionally, the enthusiasm generated must be tempered by the management board in the best interests of the city.

A smaller effort of the humanities system is to study earth's downfall and make recommendations on how it could have been avoided. In time, this work will be compiled as a report and held in trust for future generations. Included will be the multiple crises of the 70's and the 80's, the second American Depression, the Third World famines and plagues, the revolt of the Third World nations which led to their obliteration by the countries of the northern hemisphere, the squabbling among the northern allies over their newly acquired lands, the continuing increase of all forms of pollution, and the final war.

Propagation of the arts also comes under the humanities system, which puts on plays, concerts, readings and exhibits. The city's professional artists are joined by many other citizens, and the result is an unusual blend of highly professional and community theater.

All pure research is conducted under the auspices of the research system. At present, the efforts of this smallest of the city's seven systems are minimal, because there are more pressing needs. The advantage of a stable population, willing to sacrifice personal wants for the community good, is beginning to make itself felt in the research budget, which has grown steadily for the past two years.

Last of the systems is development and engineering. D and E was originally conceived by the earthside staff as the link between research and manufacturing. In actuality, the small research budget has largely prevented D and E from filling its intended function. Until now, this system has been primarily engaged in construction and maintenance, and in assisting manufacturing's craftspeople with the equipment repairs.

There are close ties between the seven systems. Life support and health both work closely on questions of nutrition, humanities and health both work on problems of mental illness, and education ties in with all the other systems. Communication is easy since copies of all documents are stored in computer memory. In Terra-Lune, the tedious preparation of formal intersystem reports and memos is only a dim memory.

Although life on the moon is not as easy as it was on earth, the people of Terra-Lune still enjoy recreational activities, especially those that lend themselves to the low lunar gravity. Ballet, for example, has taken on a new dimension as the dancers float gracefully across the wide, open-air stage. Sports are also very popular, and a sports dome was completed last year. Tennis, with the ball adjusted for the low gravity, is the favorite sport, followed by lunar versions of running the high hurdles, basketball, volleyball and wrestling.

The youngsters' favorite facility in the sports dome is the "drop tank." This tank is best described as a greatly expanded bounce shaft, used for low-gravity gymnastics. It is a cylindrical pit dug into the lunar surface to a depth of a hundred feet, and fitted with a stiff net across its fifty-foot diameter, near the bottom, for rebounding.

Although recorded music of every description is broadcast by Terra-Lune's radio station, and cathode ray tubes are used throughout the communications network for information displays, there is no television programing for entertainment purposes. Governmental and business meetings are televised for anyone interested, but those are all our budgets are likely to allow for years to come. Entertainment has returned from the electronic to the human domain, and most of us applaud the change. Friendships and family ties are stronger, and visits, dances, sports and the arts fill the time available for recreation.

Now, in the first year of the third millennium (as measured by the old earth calendar), Terra-Lune and its citizens have a firm foothold on the moon. We have high hopes that the worst is behind us, and we look forward to further improvements in our way of life, improvements that will allow us to relax a bit more and to plan for the future of our city and, eventually, others like it.

With the exception of the Returners, we seldom think of the earth, for we are too involved with life here. Memories of earth are usually much like those of grandparents who have passed away: hazy and fond, with the pleasant times overshadowing the bad. It is still too soon to tell whether the Returners will attract sufficient support to become a viable political organization. Some think not, for they would have to win over the younger generation, who have little or no memory of earth, and who therefore do not miss it. Others claim that this is the very reason why the children may be the most susceptible to the Returners' philosophy. To them, it holds the call of the unknown and the frontier, as the moon did for us. But basically, Terra-Lune is home to our children, a place that has always been, and they can imagine no other. They regard the history they learn of earth much as we do Greek mythology and the plays of Shakespeare.

Of our once "uninvited guests," little can be said. They have become so much a part of the city that there are few among us who could point them out now, only four years after they arrived. Those who chose to remain for a while at the survey camp to farm did a remarkable job. It was later suggested that the survey camp/farm be turned into a retreat for those in need of a break from city life. This idea was adopted and has proved most popular. In fact, I have been here at the retreat for a week now, enjoying the quiet as I prepared this report.

6.
The
Planet
Xeno

SHIRLEY ANN VARUGHESE

Preface: Coping with the Unpleasant Surprise

As we travel into space we will face many perils. We must realize that since we have had no experience outside of our world, the multitude of varied environments, life forms and cultures man will meet will hold many surprises. These surprises may be delightful and beneficial to mankind or extremely unpleasant and fatal. The prerequisite to dealing with unpleasant surprises is admitting their inevitability. This can be very hard to do, for this admission is tantamount to admitting ignorance.

Putting aside for a moment what we don't know, let us consider what we do know. Since we can only work with the materials at hand, logically, we must garner all our knowledge from earth and our brief excursions (either personally or by electronic proxy) to the moon and our nearest planetary neighbors.

Of every phenomenon on earth we can learn: What were the materials or elements involved, the conditions under which the phenomenon occurred, and what the results were. If you put this in an equation, it would read:

$$(Ee) + (Ce) = Pe$$

Although highly simplified, this is suited for general rather than specific observation. In the formula (Ee) equals the total elements earthly, combined with (Ce) the total conditions earthly, equaling an earthly phenomenon (Pe).

This formula can be applied to alien phenomenon with a as the alien factor involved. The equation would read:

$$[(Ea) + (Ca) = Pa] * X$$

with (Ea) equaling the total alien element, (Ca) equaling the total alien conditions, combined to make an alien phenomenon (Pa). The alien formula is governed by the X factor, the unknown aspect of the equation. It can fall anywhere in the equation and totally change the expected results. We know nothing about alien phenomena or the elements and conditions that create them. In this light, the formula appears worthless.

So why bother to present it? By using this formula to express an earthly phenomenon,[1] we can select alternatives for any or all factors in the formula based on our current chemical, physical and sociological knowledge, which may or may not be sufficient. Any change would result in a phenomenon that does not occur on earth, hence a likely candidate for an alien phenomenon.

We can propose both the possible and the probable along with the impossible. By learning to expect the unexpected, we eliminate a great deal of the X factor, even though the risk can never be negated. The formula can be applied to social phenomena as well as natural phenomena. I used the formula in the preparation of this paper, working with the assumption that here on earth, our environment played a part in every phase of man's biological and cultural evolution. Knowing this, we can try to postulate how alien environments will act on us, and possibly, how the aliens' world effects the alien, thus helping us if we should contact a nonhuman culture on one of our great adventures!

Mother Earth: Man and His Environs

The delicate environmental balance that governed the emergence of life on our planet also dictated the type of life forms that would be found here. It is presumed that the basic carbon atoms linked with oxygen, hydrogen and nitrogen to form prelife molecules.[2] These molecules formed when earth did not even vaguely resemble earth as we know it today. At that time earth's atmosphere was assumed to be composed of methane, ammonia, hydrogen and water vapor. Add to this potent radiation,[3] and given enough time, the stage for the formation of prelife molecules is set.

To express this in terms of the equation $(Ee) + (Ce) = (Pe)$:

$$(Ee) = (C + H + O + N)$$
plus
$$(Ce) = (\text{natural chemistry})$$
equals

(Pe) = (atmospheric conditions of H_2O, CO_2, CH_4, NH_3, and H_2)

Then let:

(Ee) = (H_2O, CO_2, CH_4, NH_3, and H_2)
plus

(Ce) = (radiation and natural chemistry)
equals

(Pe) = (amino acids, sugars, acids and bases) [4]

Further, given enough time, through random inter-action these basic molecules could produce complicated protein molecules, nucleic acid, cellulose and lipids, all basic prelife forms or biopolymers. In the formula: (amino acids, sugar, base and acid) plus (random natural chemical selection and two to three billion years) to equal biopolymers.

We see that under proper conditions chemical interaction has produced biopolymers.[5] To continue the process, the proximity of the biopolymers and natural chemistry would eventually lead to life.

Up to this point we see the formation of the basic one-celled life as the result of natural chemistry, the interaction between the commonest elements in the universe: C, H, N and O, plus the proper conditions (presumed to be intense radiation and time).[6]

What happened after this? Surely no life now on earth could survive in the early earth's atmosphere, or in its climate either. An organism's main weapon against extinction is its innate ability to adapt to slow changes in the biosphere, whether it be in the climate and/or the food chain. Nature selects the most suited life form via what we choose to call "Darwinian natural selection." In order to survive on the planet earth, life adapted to the environment presented to it. Every form of life on this planet reflects its total adaptation to its environment, including man.

Men have been known to physically adapt to their environments. In Peru, tribes were found living at 15,400 feet above sea level (about 3,000 feet above the zone believed inhabitable by man). It was found that they had two quarts more blood and one and a half times more hemoglobin.[7] Many generations lived at these heights before the change was completely passed on genetically.

Change in the environment necessitates adaptation. Adaptation to change is the key to survival, and survival is the main ambition of all life forms. The formula for adaptation is: (life form) + (slow environmental change) = adaptation.

The culture man builds also depends on his environment. Man has certain needs he must fulfill; he must feed himself, find shelter and reproduce. As a social creature, he needs others of his kind around him. The way he orders the process of filling his needs is called his culture. The materials he uses to fill these needs are gleaned from his environment. One can hardly build a society of fishermen in the middle of the Sahara! Man has adapted his life style to every region and every climate of the earth. As his culture becomes more technically complicated, his needs also become more complex. Again, his more complicated needs spur more technical advancement. This technical environment also breeds its own kind of cultural adaptation. A prime example is the relationship between the American and his car.

When man travels off the earth to inhabit new worlds, he will adapt his life style to suit the new environment. If there is a variation in climate, there is a strong possibility that physical adaptation will occur and be passed on genetically over a period of time. It also can be assumed that alien life forms and life styles evolve according to the requirements of their worlds.

Only when we meet the alien face to face (?) can we know what life styles and life forms abound on other worlds. Until then, we can prepare ourselves only by intelligent conjecture based on our own experiences. These conjectures form the basis for the communities both earthly and alien described in this chapter.

The Terran on Other Worlds

What types of cultures are possible in extraterrestrial communities? The answer to this question is that the possibilities are as many and as varied as the worlds man settles. Since man orders his life to fit within the limitations of his environment, whether this is a natural environment or one he has created with his technology makes little difference.

Along with the environment, the purpose of the community determines its cultural shape.

In the past, man moved across continents for many reasons. Europeans moved westward from the Old World for the thrill of exploration, to gain knowledge of strange lands. The second motive was profit. Fur traders were the first to travel to the inner regions of the North American continent in search of tradable goods for personal gain. Another reason people left their homes was to escape from an oppressive government or an oppressive economic situation, or both, in the quest for personal freedom. The prospect of free land stirred man's innate longing to have some place that irrevocably belonged to him. These reasons are the same ones that will motivate people to enter the realms of space.

These early settlers learned to live with their prairie environment. Since the vast ranges of grasslands offered no timber, the settlers learned to use dung for building houses and for use as fuel. Although the homesteaders were generally suspicious of strangers, they helped each other whenever possible, realizing how vulnerable they were on their own.

The settlers matched their laws and social order to the purposes of the community and to the dictates of the environment. Today the concept of hanging a man for horse theft is viewed as cruel and unusual punishment. But this law was specifically suited to the demands of the era. Stealing a man's horse was equal to killing him. A man stranded in the wilderness, even with supplies, was in serious trouble. The loss of a horse also deprived him of his livelihood. The farmer needed his horse for plowing, and the cowboy to herd cattle. The horse was a valuable commodity, vital to the survival of the settlers, and hence the severe penalty for its theft. In the future, as in the past, man will make laws according to the special needs arising from the demands of his new world.

Once the function of the community is established, then, with the environment in which the community is to be developed, the guidelines for cultural patterns are set. (Community function) + (Environment) = Community structure.

The two basic kinds of cultural environs to be considered when discussing extraterrestrial communities are natural environments (one in which man will not depend on extensive life-support systems for survival, such as a colony on a planet with a breathable atmosphere and a climate that is neither too hot nor too cold) and those communities totally dependent on technology for life support (such as a moon colony, or an orbital satellite).

The Problems of a Research-Oriented Community in a Technical Environment

What type of society is generated in a closed technical environment? First, let us assume that the purpose of the first moon community is purely scientific in nature, with research as the prime concern of the inhabitants' lives. Secondly, because of the lack of an atmosphere and the extremely hostile climatic conditions, technical life-support systems would be needed. This technical environment offers a very limited range of stimuli; add to this the narrow areas towards which the research scientist directs his thinking, and many problems will arise.

The inhabitants of this world will fall into two main categories: colony management and research scientists. The colony management is further divided into two subgroups: administrative, those who keep the community running smoothly; and technical, those who keep the life-support systems working.

LIST OF MANAGEMENT PERSONNEL

Administrative:

1. *Chief Administrator*—Coordinates all social and working activities.
2. *Supply Administrator*—Regulates distribution of goods and research funds, equipment, etc.
3. *Medical Personnel*—Keep the people healthy.
4. *Psychiatric Personnel*—Keep the people mentally healthy.
5. *Consulting Anthropologist*—Studies the colony and tries to waylay any impending social catastrophe.
6. *Conversion Research Officer*—Studies how to make best use of materials available in the natural environment (such as breaking down soil and recomposing it in the form of liquids, metals, oxygen, etc.) .

7. *Astronomer meteorologists*—Predicts meteor storms, solar flares and any serious climatic changes.
8. *Transportation Officers and Pilots*—Function to keep the supply lines open.

Technicians:

1. *Engineers*—Design life-support systems, make improvements and conduct major repairs.
2. *Technicians*—Operate and monitor life-support systems and keep them functioning.
3. *Communications Officers*—Keep the lines of communications open.

In addition to the management personnel, whose jobs are essential to the colony's upkeep, there are the research scientists for whom the colony was set up. These scientists will direct their studies to three basic areas. One group will center their activities on earth, making pollution studies, weather studies, resource distribution and communication studies, much in the same manner as the Skylab crews did. The second group will direct their studies toward the sky. With no thick or polluted atmosphere to distort the images of the stars, astronomers will be able to make extremely accurate computations and observations. The third group will study man's reaction to and interaction with his new environment: the anthropologists, sociologists and psychologists.

With this varied array of interests and fields of study, the community would operate more like a university, full of interdepartmental politics and intrigue, than like a colony. A cover story in the *Atlantic Monthly,* appropriately entitled "Bad Days on Mount Olympus: The Big Shoot-Out in Princeton," graphically describes the conflict in a pure research-study environment such as is found at the Institute for Advanced Study at Princeton.

Once called an "intellectual hotel, dedicated to the preservation of the good things men live by" [the institute] has been regularly shaken by internal battles of exceptional ferocity since the day it was founded . . . the strife goes far beyond the kind of healthy jousting that, for example, Einstein waged with Niels Bohr over quantum theory. Rather it has been marked by personality clashes, power struggles . . . rivalries between disciplines, and an impressive record of intellectual arrogance.[8]

The academic climate in an extraterrestrial colony would be the same, complicated by the problems inherent in a technical life-support system.

Intellectual polarization will order the social structure of the moon colony. Unfortunately, the technicians would be on the bottom of the social hierarchy. (Quite simply, how does the man who designed the computer feel about the programmer?) All of the old enmity between fields would arise. Right above the technician would be the engineer and other adherents to the "applied sciences." As the work becomes increasingly abstract, in disciplines such as physics and math, the level of social acceptance would rise proportionately. Everyone would avoid the sociologists and anthropologists in the group, basically because of man's intrinsic aversion to being "studied."

In addition to the natural polarization between the various fields of study, the inhabitants of this technical environment would be living in seclusion. Thus a basically related and pointedly dangerous situation could arise from the natural polarization between the various fields of study combined with the problems of living in a closed environment. Man does not function well in close quarters, particularly when he is incompatible with the people with whom he is living.

The solutions to both problems are related. The first step would be to minimize the feeling of confinement and as much of the technical aura as possible. One solution might be a "solarium," a greenhouse where food and oxygen-producing plants are kept. It would be a large glasslike structure where the colonists could wander without spacesuits and without the tensions arising from a closed, technically maintained environment. Building a solarium would be costly, but worth the expenditure for the colonists' emotional and psychological stability. An effort should also be made to make the dwellings as comfortable and as sensually stimulating as possible, thus widening the range of stimuli received by the settler.

Furthermore, there should be mandatory leaves outside of the technical environment, and constant communication with the outside world in the form of entertainment (films, records, books, TV, tapes, newspapers, and on rare occasions, entertainment by a troupe of actors, an interplanetary U.S.O.) . Personal communications with friends and family on the home planet, either by correspondence or radio, are very important to keep the colonists in constant contact with the larger reality outside of their secluded existence.

The second part of the solution, the elimination of the natural polarization between fields of study, is not quite as simple. The solution requires a new type of scientist who specializes in more than one discipline. An ideal engineer, for example, would have a medical degree and an advanced understanding of astronomy. He would then have an area of specialization in common with a fellow researcher who is a mathematician-astronomer-communications engineer. This not only would minimize the conflicts, but would reduce the number of personnel needed as well. All data gathered would be returned to the earthbound specialists for in-depth studies. This would eliminate the social hierarchy and the research colony could proceed with exploration.

Xeno: An Extraterrestrial Community

The proposed colony founded on a habitable world is a mining community called Xeno. It was founded with the purpose of producing the material C, which is the catalyst used in converting matter into antimatter and is vital to the matter-antimatter drive for the interstellar crafts. The value of the element C to earth's culture during an era of intergalactic exploration and colonization is undebatable. C, which was discovered in vast quantities on Xeno, has been found elsewhere only in small amounts. At first it was discovered on earth in the moors of Scotland, and later in a more active state in tropical swamps. Still, the supply was insufficient. Later exploration of Venus supplied more, but hardly enough to meet the demand. Small probes with matter-antimatter drives were launched to search the universe for this element, and once a superabundance of C was located on Xeno, man immediately set out to colonize this planet. Probes were sent in and studies were made of Xeno before the miners were sent.

PRELIMINARY REPORT ON THE PLANET XENO

Xeno is a planet revolving around an earth-type sun. There are three other planets in the system—all smaller and all orbiting closer to the sun—plus one planetoid orbiting on a vertical plane (as opposed to the horizontal plane of the four other planets in the system). Xeno's orbit is slightly elliptical, averaging 77.7 million miles from their sun. One local "year" takes 447.7 standard days (295.68 local days). The planet spins horizontally (as opposed to earth, which spins vertically), with poles to the east and west. The planet turns once on its axis every thirty-six hours (standard); thus the local day is thirty-six hours long. Twice a local year, on local days 25.16 and 273.01, the small planetoid (with a diameter of approximately 3,000 miles and an orbit of 77.7 million miles) comes within 480,000 miles of the planet Xeno. This results in a minor gravitational flux on the planet. The planet Xeno has a diameter of 11,650 miles. Although this planet is larger, it weighs less than earth (around half as dense as earth, suggesting a lack of heavy metals).

The atmospheric content of Xeno is considered breathable, consisting of 40 percent carbon dioxide, 38 percent nitrogen, 15 percent oxygen, .2 percent argon, 3 percent hydrogen, plus 3.8 percent trace elements.

The planet's climate is extremely hot. A dense cloud cover that rarely breaks (except occasionally at the poles) causes thermal inversion. Temperatures average from 85° F. to 110° F. at the poles; 110° F. to 225° F. in the "temperate zones"; and from 225° F. to 250° F. at the equator. There is a continuous rain in the "winter" (a year-round condition at the poles), and hot, steaming mists the rest of the year, with only occasional dry spells.

Life readings show an abundance of plant life, lower forms of animals, some pterodactyl-type animals and indications of "middle creatures"—half plant and half animal (possibly evolved from the *Euglena*-type one-celled animals). There are no indications of culture of any kind, either past or present.

Probe reports show the topography to be mostly marshy lowlands. The poles are again the exception, with gently rolling hills that occasionally reach mountainous proportions. There are several large lakes in the temperate zones and at the poles. Clusters of extremely large blisters, ranging from one to twenty miles in diameter, are scattered across the planet in all regions except the East and West Poles. These blisters are considered to be completely harmless plant life.

Because of the large amounts of C to be found on Xeno, and because of the urgent need for this element, it is suggested that permanent mining operations be set up on this planet. The main factor to be overcome on the planet is the extreme heat of the planet's surface. The heat can be controlled by the use of air-conditioning units wherever necessary. Enough water for the colony's use can be collected from the condensation of water vapor in the atmosphere, aside from the marshland water supply. Food can be gathered from the many life forms on the planet, and after careful studies, plant and animal hybrids can be imported from earth for use on Xeno.

It is also strongly recommended that the settlers locate their colonies at the poles, where the heat is not as intense and the load on the air-conditioning units will not be so great. There is more high ground available at the poles to build on. Later when there is more knowledge of the planet's environs, it will be possible for the miners to move closer to the equator, where the C is more plentiful.

The United World Commission accepted the recommendations as valid and sent the first settlers to the planet Xeno. Because of the importance of the C, it was decided that Xeno would remain under the UWC's control.

THE MINING COMMUNITY ON XENO

Because the function of this society is to provide the urgently needed C element for use in the interstellar space drives, all of Xeno's work forces are maximizing their efforts to produce it.

Here is a brief outline of services necessary for the production of the element and maintenance of the community:

I. Mining Operations
   A. Engineers
   B. Technicians
   C. Miners
II. Base Operations
   A. Environmental technicians for climate control
   B. Food-production managers
      1. Production and preparation
      2. Distribution
   C. Planetary communications engineers and technicians
   D. Planetary transportation engineers and technicians
   E. Housing "engineers" and builders
   F. Meteorologists
III. Individual duties aside from mining operations
   A. Health services
      1. Mental health
      2. Medical care
   B. Child care
   C. Education
   D. Security patrols
      1. Settlement security
      2. Mining security
      3. Planetary security

      E. General services
          1. Operate shops
          2. Cleaning services
          3. Secretarial and clerical
   IV. Researchers (to add to the colonists' knowledge
       of the planet)
      A. Exobiologist
      B. Exobotanist
      C. Topographer
      D. Anthropologists and exoanthropologists
      E. Linguists and translation experts
      F. Astronomer

THE STATUS OF RESEARCH IN A WORKING SOCIETY

Unlike a society organized solely for research, this is a working culture, oriented toward hard labor. Since the engineers and technicians are the essential personnel needed for the accomplishment of the community purpose (C production), they hold the highest positions in the social order.

Intensive scientific research is not the main objective of this community. The settlers are too concerned with the production of C and with simply surviving in their hostile environment to bother with in-depth studies. Intensive research would be treated as purely recreational. On the other hand, basic research is a necessity. For example, the first settlers are more interested in knowing the best way to cope with the fact of the carnivorous plant, than in studying the plant's morphology. Vital knowledge includes animal behavior, analysis for edibility, and the animal's function in the biosphere.

THE FAMILY: AN ALTERNATIVE

The pressures of C production plus the harshness of the environment produces a spartan society. Skipping work assignments results in social ostracism or deportation.

The severity of the work and the hardships of the planet make recruiting personnel difficult, if not impossible. Due to the distances and the difficulties involved in recruiting personnel, the main source of population replacement is reproduction. The need to replace the population is directly related to the need to establish a permanent colony for the mining and production of the C element.

Since some social orders are better suited to certain objectives (in this case, rapid reproduction for the purpose of population replacement), a community organizer makes his plans accordingly.

A social order based on the need to produce the maximum population growth possible from the small number of settlers available calls for a larger number of females than males, at a suggested ratio of five to one.

This group of five women and one man, plus offspring, is protected by law, as is our current family. But in this society every member of the family over the age of fourteen is required to work on behalf of the family. Young males are encouraged to leave the family group as soon as they reach the age of sixteen. Until they reach eighteen, the age of consent, they live in hostels specially designated for the purpose of housing unattached males. The daughters have the choice of staying with the group and adding another male for every five female family members that come of age; or forming her own family, with females with whom she feels more compatible. The selection of family members, according to law, must be done with the written consent of all members involved. Hence, the marriage license has six signatures instead of two. (Although six is the ideal number for a family, legally the number could range from a minimum of four, three females to one male, to seven, one male to six females.) Any more or less is considered impractical and definitely unacceptable.

This arrangement offers a definite advantage to the female members of the family of childbearing age. In a monogamous society under the same conditions it would be socially acceptable but medically unsound for her to produce one child a year. This not only recklessly endangers her health, but it reduces her capacity to work productively on other physical and mental creative efforts. In this system births are staggered to allow each woman time between children and longer periods of "usefulness" to the community.

Of course any social order is liable to change. This social order will remain stable as long as there are more women than men. Overpopulation might also endanger this social order and force people to return to monogamy. If the settlers were satisfied with the social order, they would employ methods of birth control rather than social change to control the population.

The main danger in this society is that the minority of males might become socially oppressed. (Anywhere there is a minority this is a real possibility.) The only way to combat this or any other type of oppression is to nurture respect for the minority. Since the Xenophian society is based on a work ethic, respect is gained from the ability to do one's job well. All social judgments are based on this fact. If work loads from mining to nursery duty are allocated equally, and performed equally well by every member of the community, then a feeling of female supremacy will be minimized.

All members of the family care for the children. Each child receives an equal amount of love and care from each one of the mothers, whether she is the natural mother or not.

Education is handled on an individual basis by tutors. The tutor is anyone in the mining community who has something to teach, whether it be basic math or music appreciation. Some classes are conducted on more technical materials that require more advanced knowledge. Children are taught the various community jobs by direct participation and a minimum of theory, and gain a sense of purpose early in life.

Because of the C production pressures, it is imperative that the mining equipment be kept running thirty-six hours a (local) day. During most of the year, it is impossible to keep the temperature of the mining areas under 100° F., and since the community must keep production at its peak, they must rotate mining machine operators every hour. Thus every person over the age of fourteen must work in the mining fields. All outdoor jobs must be rotated for the same reasons; therefore everyone needs to know more than one job.

GOVERNMENT AND LAW

The government of the mining community is a mixture of democratic capitalism and socialism. It is a cooperative system in which government and business are one and the same. Mining is the only business on the planet and all of the mining company's net profits are directed back to the colony. Each family receives allotments for food, shelter, clothing and city services, according to each family's needs. The company's net profits are equally divided among the populace. Monetary bonuses and work leaves are awarded on the basis of "creative thinking" for the improvement of life on the planet.

The head of the government is the president, an elected official directly responsible to the people and in charge of business and community affairs. Below the president are two more elected officials, the vice-president of business and the vice-president of community affairs. Their duties extend into the areas designated by their titles.

The president appoints his staff of secretaries, clerks and advisers. There is only one position that he appoints that must be approved by both the vice-president of business and the vice-president of community affairs, and that is the vice-president of work distribution. The president nominates a man to fill this position and the two vice-presidents vote on him. The vice-president of work distribution assigns people to their jobs, and coordinates work assignments. It is important that this man be impartial, for in a community where work is the main function of each member, any partiality would be met with radical hostility. It is the president's job to see that the work assignments are fair.

The vice-president of business also appoints his staff. Besides secretaries and clerks are the chief of negotiations, the chief of personnel, and the finance and administrative officer. The chief of negotiations barters with the government officials of the United World Commission for the price of the C and for other non-Xeno government trade and tax concessions. The chief of personnel selects men for the managerial posts in the business part of the government. These men are selected on their merit as managers, and are approved by the vice-president of business affairs. The finance and administrative officer makes recommendations as to the best way to increase profits. He handles all the administrative aspects of the business.

On a level with the vice-president's staff are the members of the line management structure, or the chain of command in the business portion of the government operations. They are the chief of base operations, the chief of exploration and mining research, and the chief of mining. The chief of base operations has under his jurisdiction all aspects and functions necessary to maintain the base for mining operations: climate control, food production, communications, transportation, housing and weather reporting.

The chief of exploration and mining research conducts surveys to locate new C fields and researches better methods of producing the element C. The mining chief is in charge of the actual mining operations, maintenance of the mining equipment, and all mining engineers and technicians.

The second half of the government is run by the vice-president of community life. His staff includes the health chief, in charge of the physical and mental health of the settler; the finance and distribution chief, in charge of managing the settlement's distribution of goods, housing, services and excess funds, and the director of education and child care. The vice-president of community life is responsible to the people's representatives, elected by the people. They serve as intermediaries between the community and the vice-president of community life. These representatives are responsible to the people who elected them.

The laws regarding stealing, murder, etc., are the

basic laws of mankind. Local laws would include special penalties for neglecting work assignments, smuggling C, and endangering the life of a colonist or of the entire colony. Punishment would not be severe and the main emphasis would be on the prevention and cure of crimes.

MAN AND ENVIRONMENT ON XENO

The climate of perpetual rains and intense heat will affect the way the settler builds his homes and plans his physical community.

The homes built on Xeno are either A-frame or dome to prevent water pressure from rain accumulation from causing leaks and stresses. Air conditioning is a necessity due to the extreme heat on the planet's surface. The need for illumination due to the planet's perpetual twilight is met through extensive lighting and an abundance of windows, which do not open. There are also tight seals on every door to keep the cool air in and the hot air and constant downpour out. Because of the lack of metals on the planet, practically everything is constructed of wood or plastics.

Since the planet is a swampy bog, the community is built on the highest, dryest ground available or on stilts. Because it is a mining community, it is built close to the mining fields. (On Xeno the C is found in the swamp silt; the silt is mined, and the C is removed.) The distribution of high ground is a matter of priorities. The main concern is the protection of the mined element C and the mining equipment and supplies, so these are stored first on the dryest location possible. The community life-support systems and the water condensation and purification plants, the community air-conditioning plant, the power plant and food-supply warehouse are also distributed on the highest ground. After all the essentials are located, the colonists have any of the land left.

The Aliens

In their rush to mine the C the Xeno settlers failed to study the oddities of the planet. The preliminary report mentioned "extremely large green blisters, from one to twenty miles in diameter, scattered across the planet; found in all regions except the poles," but since the scanner analysis showed that these blisters had chlorophyll, it was assumed that they were merely local plant life. Since the settlers rarely left the polar regions, as it was too hot, and there was no need to explore beyond the polar C fields while they still produced, it was not until the settlers had survived for seventy-five local years that they noted the peculiarities of the blisters.

While moving through the heavy jungle swamps a prospecting team working near one of the smaller blisters, about half a mile wide, saw a humanoid-type figure dive into a rift in the surface of the blister. The rift then sealed immediately.

The news of this occurrence disturbed the mining community, because they realized how ignorant they were of the planet as a whole and the blisters in particular. Studies were instituted and preliminary reports indicated that there were indeed humanoid life forms living in what still appeared to be large plants. It was not until the colonists tried to enter the blister that they encountered trouble. They found that they had trespassed on an alien city, a culture that they had failed to discover because they had no equipment to detect a technology based on nonmetals, such as plastics, and rubber, or even to detect a superior life form that was more plant than animal.

If the aliens had been hostile, the settlers would have been in serious trouble long before the trespassing incident. The native Xenophians had been provoked to action when the miners tried to invade their city—that is, enter the "blister." Even then the Xenophians merely apprehended and held the miners. Due to the Xenophians' innate curiosity and their protective coloring, they had studied the miners undetected since the first settlers stepped off the ship.

Once the initial contact had been made, a careful cultural study was made. The colonist also attempted to reimburse the Xenophian for the use of his planet and for the valuable C they had extracted, but the Xenophians could not understand the worth of the element, nor the meaning of rent, nor the method of payment. In terms of social values, the miners and the Xenophians were a mystery to each other.

THE XENOPHIAN

Physically, the Xenophian stands between seven and nine feet tall and weighs from 150 to 250 pounds. The Xenophian has from one to three sets of tentacles on his upper abdomen, all of which he can manipulate with dexterity together or separately. He walks upright, on two muscular legs and two webbed feet. He has one set of auditory sensors and one set of infrared sensors mounted on flexible "antennae" on his forehead. Due to the twilight conditions of the planet created by the cloud cover, he has large, dark, nocturnal eyes located mid-head, which are designed to admit as much light as possible.

The complexions of the native Xenophian range in shades of green to blue-green. The Xenophians utilize chlorophyll, deriving most of their nourishment from photosynthesis. Unlike rooted plants, they gather nitrogen by ingesting smaller animals. The Xenophian needs a small protein meal every three or four local days. The food is ingested directly into the stomach cavity from a mouth located in the center of the Xenophian chest. The digestion consists of breaking down the animal proteins to basic nitrogen, which is then transported to the cells by a circulatory system much like our own. Anything that is not nitrogen is expelled directly from the stomach after the digestive process is complete. Liquid and other waste products from the cells are expelled through the pores of the skin, as he has no lungs and no waste-removal system. He has six to eight simple hearts located in various parts of his body.

His main need is for moisture. He drinks only one or two ounces of water with each meal to aid his unique digestive process, and absorbs the rest of his liquid directly from the very humid atmosphere. During dry spells the Xenophian is driven into a dormant state. He forms a spore and remains in estivation until conditions become favorable again. Once it starts to rain, he assumes his former size and shape.

The Xenophian is monoecious; he has both male and female sex organs, or pistillate and staminate flowers. There are two stages in the Xenophian life cycle. The first is spent in the form of a rooted plant. This plant's fruit carries a fetal Xenophian that is expelled when the Xenophian reaches maturation. At that point the Xenophian grows to maturity as would any animal. Once he reaches the age of sixteen local years he is capable of reproducing. This entails pollination of the flowers that blossom on his head. Once the seeds have matured, they are planted and produce the rooted plant stage of Xenophian life.

The Xenophian has developed a complicated symbiotic relationship with a smaller plantimal they call "the helper." In primitive times this creature performed the function of pollinating the Xenophian flowers. Today, the helper is still socially and biologically important to the Xenophian.

The Xenophian protects himself from predators with his intelligence, his protective coloring and his excellent reflexes. However, the Xenophian is severely handicapped by his inability to function after sunset. Photosynthesis stops and the Xenophian is unable to store enough "energy" to conduct any activity but the basic life functions. This is the time that he is most vulnerable to the planetary predators. If the Xenophian is outside of the dome or "blister city" when the sun sets, he may as well resign himself to the eventualities.

Because the Xenophian falls prey to the larger animals, and because his natural life span averages only thirty-five local years, it would be very hard for the Xenophian to make any social or technical progress if it were not for his ability to pass on knowledge genetically. This knowledge includes what the Xenophian has learned in his own lifetime, as well as the knowledge he has inherited.

XENOPHIAN TECHNOLOGY

By all of our standards, the lack of metals on Xeno would have eliminated the possibility of an advanced technology. But the Xenophian has developed a superior technology with the materials available to him. Controlled fire came into use very late in Xenophian development. The misty, liquid climate made it hard to sustain combustion, and when fire was finally put to use, chemicals rather than wood and other combustibles were employed. Fire as a source of warmth was impractical and unnecessary in the extremely hot climate, and any heat that was needed for industrial uses or food processing could be supplied through the numerous hot springs. If the Xenophian wanted to cook his occasional meal, for example, he would drop the food into the superheated water of the hot springs.

Outside of primitive tools and weapons, the Xenophian's first major discovery was rubber production. Later he developed polymers and plastics that formed the basis of his technology. The discovery of crude rubber came easily. A Xenophian noticed how the "sap" of some plants near the springs reacted when it dropped into the hot water.

Before the development of the rubber polymers and plastics, the Xenophian fortified himself against the planet's many predators by using the available materials to build walls out of wood; and nets that stretched across the top of the compound from vines. The blister responded like a membrane, selectively admitting light and moisture and expelling oxygen. The outer layer of the blister could repair itself unless damage was too serious, and then repairs were made in the same manner as a skin graft. The blister aided in moisture condensation and kept the Xenophian functioning in the dry seasons.

Because of the hardships of traveling long distances through the shallow cluttered swamps, the Xenophian had few contacts with members of his race in other blisters until they developed the balloon, a natural result of their knowledge of rubber and the abundance of natural helium escaping from gas pockets in the swamps. The first Xenophian aeronauts lifted off about five millennia earlier in their development than did man.

XENOPHIAN SOCIETY

Since the Xenophians are monoecious and do not need another member of their race to reproduce, they are not likely to form a family unit as such. But the Xenophian is so vulnerable to the many hazards of the planet that he has aligned himself with other members of his species. The Xenophian society is based on the concept of group survival. An individual realizes that he can only survive if his group as a whole survives in the present and in future generations. Thus the ultimate antisocial act is endangering the nursery—that is, posterity.

In this social order, crimes are few. Stealing from the Xenophian is impossible because he has few possessions, and none that he would not give away. The Xenophian vocabulary does not even contain the words "yours," "mine" or "his."

The Xenophian's strongest emotion is fear. He does not love; at most he feels fondness and friendship. The most negative emotion a Xenophian can feel is disgust or dislike, but he cannot feel hate.

Sterilization is the harshest punishment that exists. The Xenophian sees his seed-spores as his personal bid for immortality. If the Xenophian wishes, he can reproduce another being exactly like himself, physically and mentally. With hereditary knowledge, the new Xenophian literally knows what his progenitor knows. The parent Xenophian feels, even more than an earthling ever could, that he is "continued" in his offspring. Sterilization is equivalent to excommunication for the Xenophian.

This is not to say that the new Xenophian is merely a carbon copy of his parent. He exists as an individual with his own identity and experiences. The hereditary knowledge is not passed on as memories per se, but as ready reference materials. For example, the progenitor's encounter with a frightening, giant praying mantis is passed on to the progeny in the form of knowledge that the praying mantis is extremely dangerous, and that shoving a pointed stick up through its mouth will kill it. The details of the struggle with the mantis are not included in the knowledge transference.

The average age of Xenophians who die a natural death is thirty-five local years, not including the time the Xenophian spent in estivation (or in the dormant spore state). The average age of the Xenophian—including those who have died as victims of the elements or of predators—is twenty local years.

In the Xenophian society there is a process called "The Telling," in which the newly matured Xenophian, even before leaving the nursery, tells everything he knows to the "corrector of knowledge." It is the duty of the corrector to correct any false knowledge that might have been passed on by his progenitor. If the new Xenophian tells the corrector something that conflicts with established fact, the corrector researches the information. If the information proves to be correct, then the new Xenophian is instructed to retain the knowledge, and the corrector adds the new information to his stores of knowledge. If the knowledge is proved to be false, then the new Xenophian is instructed to discard the information. The position of the corrector is hereditary.

By planning genetically over several generations, the Xenophian has produced leaders with a superior capacity to rule. Hereditary knowledge, special capabilities in economics and law, and an interest in the success of the "group" were carefully bred into the leaders. The most carefully governed genetic planning went into the production of the "correctors of knowledge," who needed to know virtually everything. The correctors studied their own hereditary knowledge and learned how to simulate the process of passing knowledge on chemically, by using pollen to transfer information. For example, a group on the east side of the hot springs could exchange the knowledge of a new rubber process, in the form of pollen, with a west-side group's nursery fertilizer formula, also in the form of pollen.

THE XENOPHIAN GOVERNMENT

At the head of every group of Xenophians is the organizer, who functions as the administrator. Equal to him in status but not in power is the "corrector." Another official post is the "genetic counselor," who oversees selective crossbreeding to distribute certain mental and physical attributes among the new Xenophians; a second is the "security officer," who provides the best means for group protection and makes special security arrangements for the protection of the nursery. Three groups, each having from thirty to a hundred and twenty members, together form a clan. The officers of the clan taken from the three groups are in charge of operations for matters that involve the clan as a whole. Above the clan is the state, formed from three clans. Above the state is the empire, formed from all the states on the planet. The state officers are bound together in three separate senates according to their various functions. Above the senate is the council, made up of the three chief advisers tó the "supreme Xenophian," who is above all.

THE HELPER

The helper is a small plant-animal that looks like a small winged monkey, measuring approximately one to one and a half feet long and weighing from five to ten pounds, with a wingspan of twenty to twenty-six inches.

The helper is an intelligent creature that aided in the cross-pollination before the Xenophians assigned the task to a genetic counselor.

Currently, the helper's function is to pollinate the urban Xenophian's botanical gardens. The honey the helper makes from the flowers that bloom on the Xenophian is a vital nutrient for the fetal Xenophian. The honey is also essential to the helper. It is the food they feed to the embryonic helpers to produce fertile females.

The process of gathering pollen and nectar from the Xenophian is a very personal and private act to the Xenophian. Some of the strongest emotional ties a Xenophian forms are with his helper. There is no social function or ceremonial event at which a helper is not present.

The Xenophian considers the helper to be equal to himself, despite the differences in physical appearance and social order. The helper's social order resembles that of earthly bees, with neuter workers, male drones and female queens. The helpers and the Xenophians communicate very well in a language that consists of whistles, squeaks and gurgles.

Dealing with Aliens

As man expands his environment to include the realms of space, his chances of contacting nonhuman cultures of extraterrestrial origin increase dramatically. Thought of such contact excites the imagination and at the same time poses some very real problems.

How should we react to meeting alien life forms? The only logical answer should be "with extreme caution." When man meets an alien for the first time he will have to overcome a severe psychological handicap: his tendency to judge aliens by earthly standards.

Every individual has some preconceived ideas as to how an extraterrestrial life form should behave. We assume that the alien will be either a malevolent conqueror or a savior. We equate advanced technology with advanced social development, thinking the two go hand in hand when they often do not. We should become aware of these prejudices and try to cope with them. We should be able to recognize intelligence and good intentions, despite the outward appearance of the life form.

Alien societies can be graded according to their level of social and technical advancement by considering this advancement in relation to the amount of time it took to develop. If we were to measure the rate of scientific and technical advancement on earth, assigning zero to the time when the first man-ape picked up a rock and used it as a tool and ten to the time when he developed "controlled" fire, and then consecutively number each development of importance thereafter, we could then divide this figure by the amount of time it took to reach that level of technology. This would be the science and technology quotient of earth. This formula could then be applied to the timetable of social developments to compute the social quotient of the culture. We could therefore determine alien social and technical advancement in the same manner.

The ratio of the aliens' science and technology quotient to their social quotient $\left(\dfrac{STQ}{SQ}\right)$ will present a fairly accurate picture of their culture. If their STQ is greater than ours and their SQ is less than ours, we should be wary of making any sort of contact with that alien culture.

If we are to cope with any alien culture we must be prepared to relinquish all of our preconceptions. Anything the Terran takes for granted about his visitor could prove fatal. To assume, for example, that the alien is hostile and to treat him as such could force the alien into a hostile reaction. If the alien had superior or even equal forces, the results could be devastating. To treat a hostile visitor as friendly could be equally catastrophic. We do not even know what would constitute a hostile or friendly act to an alien visitor.

There are two types of possible encounters: the planned encounter, or the meeting, prearranged by way of radio or television contact; or the surprise meeting, the more likely and the more dangerous of the two.

THE MEETING

An earth-alien meeting should take place in an earth orbital satellite. One-half of the satellite could be equipped to maintain an environment in which the aliens could function, while the other half, separated by a glasslike wall, could contain an earth atmosphere. Communication and translation devices could be set up between compartments. Since the meeting would be conducted off earth, this would be one way to prevent social and biological contamination.

The aliens should be warned in a carefully worded "statement of hostile acts" as to what we consider a hostile act; and informed that retaliation would be the immediate result of any of the activities described in the statement. The aliens should supply us with a statement of what they consider inappropriate or hostile, along with what they consider appropriate and friendly.

Preliminary negotiations would be lengthy and expected by both parties. It is unlikely that any intelligent culture would intrude on another culture without first making a preliminary study to determine if there were any potential physical or attitudinal threats. These cultural studies would be needed to cope with the rudiments of social interaction.

It is safe to assume that the first steps to contact would be listening for radio or TV signals, such as we have done with project Ozma. If any signals were detected, they would be followed up first with probes (electronic snoopers), to see if further study is warranted. After that, research teams would be sent to the prospective contact planet. The close range studies would be made from the outer atmosphere, with occasional trips to the planet's surface.

The researchers would try to determine if there had been previous extraplanetary contact, how it had been made, and the effects it had had on the alien society. Perhaps after a careful study, the researchers would find the prospective contactees completely opposed to extra-world contacts of any kind; or they might judge the society too unstable for contact, or a definite threat to our society and would withdraw without making contact. If they found the culture safe and ready for contact, the researchers would try to find a compatible means of communication and the best agency to contact.

There are four reasons for contact. The first and most benevolent reason is cultural exchange, through which both cultures will benefit. A second and more frightening reason is that the alien intends to conquer and colonize, and we have no choice. The third reason is economic: we have something they want, or vice versa. And the fourth reason is political: either we are strategically located, and they want us to join their union, or they want us to keep out of their territory.

Whatever their reason, we should assume that the contacting culture will try to persuade us that their ideas and intentions will be to our advantage. Even if the contactors are benevolent, we should adopt an attitude of caution.

If the contactors are benign, it will be evident by their willingness to comply with the contactees' suggestions for conducting the meetings and cultural exchanges. Undoubtedly from their studies and from our "statement of hostile acts," they will know what we consider unreasonable. If it is their intent to take our world, they may or may not state their purpose before they make their attempt to conquer. The possibility of hostile aliens cannot be overlooked. We have early-warning systems to tell us of the approach of missiles from other countries, yet we have no early-warning devices to protect our planet. In any dealings with aliens we should be prepared to defend ourselves.

If contact is for economic reasons, the main question to consider is whether they will offer as payment something totally without value in their world or in intergalactic trade. If we desire reciprocal trade, we should know what constitutes the medium of exchange. Consider the American Indians who were given glass beads in exchange for their property. At the time, glass beads were of value to the Indians, but later, when they needed a firm basis for trade with the white man, they had nothing with which to trade.

If the contact is politically motivated, we should be aware of the laws that govern the alien realm. We should listen to the alien contactors for information on their regulations. Perhaps the contactors are members of a federation and would like us to become a part of it; perhaps they want us to join them the way the Southern planters wanted the Africans to join their plantations. The best course of action is not to respond until representatives can be sent to the alien federation and studies can be made.

THE SURPRISE VISIT

The surprise meeting leaves us at a disadvantage. There are three possible reasons for the surprise encounter: the accidental crash landing, being caught on a reconnaissance mission, or overt hostility.

If the visitor is not overtly hostile, the situation should be handled as follows:

1) Make *no* move in their direction, but retreat to a safe distance and wait.

2) If the aliens make a move toward you, move backward to show them that you aren't overly anxious to have them close to you. Fast movements should not be made unless absolutely necessary, so as not to startle the visitor. The best retreat is a slow one backwards so you can keep your eyes on the visitor.

3) If you have a weapon or weaponlike thing,

particularly a flash camera, lay it aside. (Consider the effects a flash of light would have on an alien who is used to a laser-type yeapon.) Laying aside your weapon would be considered friendly. If you don't have the confidence to lay it aside, lay it across your chest or let it hang down to your side, ready but not aimed. This would be a clear message to anyone who did not speak our language.

If the craft appears to be in trouble, and if the occupants appear to need special life-support systems, we should cautiously try to offer assistance. Offering could be done by means of body language to try to make our helpful intentions clear.

In no case should we try to communicate with the aliens until they have gone through proper communication channels—that is, setting up a meeting and stating their intent. The only contact which should be made with the surprise visitor is to tell the alien to contact the proper agency. At this point there is no agency set up for the express purpose of dealing with an alien landing on our world, a definite lack of foresight on the part of mankind.

Notes

1. George Gaylord Simpson, "The Non-Prevalence of Humanoids," *Science* 143, no. 3608 (1964) : 769–775.
2. Tom Allen, *The Quest: A Report on Extraterrestrial Life* (New York: Chilton Books, 1965) , p. 57.
3. Ibid.
4. Norman Horowitz, "Is There Life on Other Planets?" *Engineering and Science Magazine,* California Institute of Technology, Pasadena, California, 1961.
5. Melvin Calvin, "Chemical Evolution," Condon Lecture Publication of the University of Oregon Press. Reprinted in *Interstellar Communication,* ed. A. G. W.

Cameron (New York: W. A. Benjamin, Inc., 1963).

6. Carl Sagan, "On the Origin and Planetary Distribution of Life," *Radiation Research* 15, no. 2 (1961).

7. Allen, op. cit., p. 99.

8. Robert Manning, "Bad Days on Mount Olympus: The Big Shoot-Out in Princeton," *Atlantic Monthly*, February 1974, p. 38.

7.
The Contact Group and a Nonhuman Extraterrestrial Culture

MARY OBERTHUR

In the last few years the notion that we will some day contact intelligent beings from another planet has suddenly become believable. What used to be considered material for fiction writers is today, because of progress in various technologies, an increasingly realistic possibility. Advances in physical science enable man to leave earth and enter outer space. Advances in administration enable man to plan not only for foreseeable contingencies, but for unforeseeable ones as well. These new skills have changed our attitude: We can go out there. We can plan for the unpredictable.

Some scientists and statesmen who in the past have studied the effects of contact between unfamiliar cultures feel that we must prepare for this next possible contact. They seek the answers to these questions: When and how will it happen? How will people react to contact? What effect will it have on us? Do we have any trained persons with useful knowledge and skills in this area? What information is available or can be researched? What can we do to control the outcome? To explore some of the possible answers to these questions I have set up a fictional contact situation between an exploration expedition and the Marsupians, a nonhuman community residing on an earth-type planet which circles another sun.

The leading question is, When and how will it happen? Although I have placed our contact outside our solar system, we have no way of predicting whether we will be the Visitors or the Visited. The calculations involve too many unknown factors, but as our technology progresses, the possibility increases of our being the Visitors. How long will it be before we can go beyond the limits of our solar system? Predictions of scientific progress are often quickly outdated by breakthroughs in theory and technique. Just before World War II, Isaac Asimov, knowing that current scientific predictions were that man would not achieve space flight before the year 2000, nonetheless set a story about a "first moon landing" in the 1980's because he wished it to fall within the lifetime of his readers. This knowledgeable scientist and writer was almost right, but only by accident. At the moment scientists need a breakthrough in theory before we can escape our solar system. It may come tomorrow, or never.

For present purposes, it is assumed that when we

leave our galaxy to explore the universe, we will visit stars like our own sun. These most likely will have planets with intelligent life forms and plants like those we find on earth. If we were to land on planets completely unlike our own, we might not even recognize the intelligent life forms that have developed. An intelligent life form adapting to a Mars-type environment might look like a lichen; on a Jupiter-type planet, it might resemble a crystal. The giant planets would be avoided because of their tremendous gravity and the barren planets because of their lack of usable air or water.

The next question is: How will people react to contact? I have placed contact away from earth because we can better evaluate the potential situation by using a limited number of carefully chosen humans. If contact were made on earth, the hysteria and general hubbub that would break out would make it impossible for future historians to untangle the facts. We have seen the effects of our almost instantaneous communications system on human events. Action and reaction in news interpretation and public opinion often move so quickly that we wish for a breathing space, a chance to evaluate the situation.

Two human characteristics will make contact difficult for us. First, although man no longer needs to defend a certain number of acres to ensure food for his family, our territorial feelings are still strong. Many of our laws involve ownership rights. Neighbors quarrel bitterly over a few inches of land and nations war over a few miles of territory. Few of us welcome strangers with enthusiasm. Second, our egocentric view of the universe persists despite our intellectual acceptance of ever-larger views of reality. We have learned that the earth is not the center of the universe but revolves around the sun, and then that the sun is but one small star at the edge of one galaxy; yet emotionally we are still us-centered. Who has seen the photo of a jewel-like earth floating against the depths of space and has not felt joyous pride?

The third question is: What effect will contact have on us? It is naïve to suppose that either the humans or nonhumans will be thinking in any other terms than "us" and "not us." Intelligence presupposes the ability to discern differences and similarities. The differences will be easy to see. The similarities will be our only hope for mutual recognition and respect, a hope that will be dashed if we meet beings who do not recognize that we are intelligent creatures, or on the other hand, if we do not recognize that they are.

An example of the effects of first contact can be seen in a historic incident. Both the Spaniards and the Indians in sixteenth-century Mexico were affected by the attitude of the Conquistadores. The Aztecs were wiped out and the Spaniards were brutalized. Cortez had a problem with a priest who insisted that the Indians were human, not animals. The priest finally traveled to Rome and presented the facts to the Pope. The church took the stand that these creatures were human and therefore possessed souls. This was an important step for mankind, but one we tend to forget. We will find out just how much our attitudes have changed when we meet intelligent beings with different shapes than ours.

Planning for the possibility of future contact will be like a community formulating a disaster plan. Disaster may never strike, but if it does, the difference between emergency and catastrophe may be determined by the scope and flexibility of the plan. We have only to read our history to see that when acculturation is unexpected, unplanned and unorganized, tragedy often follows.

The meeting of cultures is not new, and for the most part such meetings have resulted in death, destruction and disease exchange. It took Europe a thousand years to recover from the loss of law and knowledge when Rome fell. In America in 1837 within a few weeks smallpox brought by white settlers reduced the Mandan tribe of the Northwest Plains from sixteen hundred people to thirty-one survivors who were absorbed by neighboring tribes.

In meeting nonhuman beings we will be guarding against the development of hostility and the exchange of harmful customs, but because of biological dissimilarity, disease will probably not be a problem. Trained persons with useful knowledge and skills must smooth the way for contact with strange cultures. The contact team will be made up of several scientists, with an anthropologist as the leader. He will have the best background to identify and evaluate the contact problems, compile the data and make recommendations.

Many anthropologists have already demonstrated their ability to ease social misunderstandings and conflict. During World War II the Japanese attitudes and customs were so alien by Western standards that anthropologist Ruth Benedict and a committee of leading anthropologists were assigned to study their culture by the Office of War Administration. One crucial question was the terms of surrender. They recommended that the emperor be retained since his leadership was closer to that of a Polynesian sacred chief than a Western dictator. This recommendation saved thousands of lives, for after the emperor's broadcast, the soldiers laid down their arms and the surprised Americans were greeted by a cheerful, cooperative people.

Thus anthropologists have used their knowledge and skills to solve new problems. George Santayana has written, "Those who cannot remember the past are condemned to repeat it." Mankind, venturing toward the stars, will have a tremendous body of fact and experience to call upon. There is no reason for us to repeat our past mistakes.

Each time man has met a barrier to progress, he has overcome it. At this moment the limitations imposed by the speed of light keep man bound to our solar system. But once we laughed at the alchemists who tried to transmute base metals into gold and today, albeit at an enormous cost, a cyclotron can convert mercury to gold. A few years ago scientists pointed out that ray guns were an impossible figment of a science fiction writer's imagination, but now they are using laser, maser, and plasma guns in their laboratories. We were once told that man could not break the sound barrier or escape earth's gravity alive, but then we experienced *Apollo 7.*

It is not unusual for man to discover new and strange life forms. Captain James Cook named Botany Bay for its unusual plant life, and today we know that the majority of the 15,000 plant species in Australia are found nowhere else. Marsupials, long thought extinct except for a few opossums in the Americas, are actually thriving. For years European biologists thought the descriptions of Australian animals that explorers brought back were alcoholic visions. When monotremes were discovered, scientists took a look at the duckbilled platypus and published papers explaining that the animal's unusual but continuous skin was a hoax.

The white settlers classed the aborigines as another weird animal, subhuman at best. When they felt there was an overpopulation they used the same technique that we use with coyotes: food laced with arsenic. The aborigines, although generally considered primitive, have the most elaborate and rigid kinship and marriage rules on earth, and the Europeans found this hard to accept. When a white man mated with a native woman he found himself accepted into her brother's family, where the brother's wives were also available as mates. However, all his wife's female relatives were taboo, and the penalty for transgressing the taboo was death. A law-abiding Australian tribesman would not dream of living in a hut with his mother, mother-in-law or adult sister because of social and sexual taboos. They observed the settlers' family patterns and were shocked by the white man's promiscuity and the fact that he practiced group marriage.

Records of man's perception of first contact as miraculous, as godly visitation, fill whole libraries. The Aztecs in Mexico had a strong and warlike civilization, yet because one of their myths foretold the coming of a white god bringing great gifts, they reacted to the Spaniards with adoration and were easily conquered. On the other hand, an extraterrestrial anthropologist visiting earth who found that various features of his race were associated in human culture with the devil, vampires, cobras, or gorgons would hesitate to recommend open, public contact without careful preparation.

Progress in biological research significantly expands our acceptance of the strange and the new. The old dichotomy of plant and animal has long fallen short in explaining organisms like *Euglena,* a group of animals with chlorophyll, and the fungi, a group of plants without chlorophyll, that would not fit under either category, and students today learn that there are four kingdoms: plants, animals, Protista (fungi and single cells with organized nucleus) and Monera (simple cells, bacteria and viruses). We once thought that cells were merely bags of jelly with a nucleus and vacuoles, but the electron microscope has shown us that cells are tremendously complicated and well-organized units of life. More and more it is found that the seemingly fantastic may be true.

Another way in which biology will help us in extraterrestrial exploration is by providing an understanding of the breadth of our biological rules on earth, and thereby enabling us to better predict which life forms could survive on an earth-type planet. Earth provides many surprising biological exceptions. There is life in areas as diverse as Antarctica, Death Valley, the Amazon Basin, and the Himalayas. Scientists are already attempting to determine on the basis of our present biological and cosmological knowledge whether there are other worlds where man can survive or where human life may even be flourishing.[1] For instance, xenobiologists, investigating which life forms could tolerate life on Mars, found that reindeer lichen (a fungus and algae partnership) and botulinus (anaerobic and poisonous) could live and multiply under that planet's conditions.

It is assumed that the goals of the exploration expedition will be research and, if possible, peaceful contact with any extraterrestrial intelligence (XTI). The expedition team will be instructed in methods of observation and communication and will have an extensive microfilm library of resource material. Assuming that the initial observations and confrontations are successful and the members of the team are accepted as visitors, they will be able to collect invaluable information about their hosts and on the interactions between the humans and the non-humans.

One of the controllable factors in contact is the actions of the humans, those on the expedition team and those back home. The team will consist of a limited number of women and men chosen for their professional competence, emotional stability and reliability of judgment. Since successful contact will be a desired goal, they will be willing to follow instruction and use care in their contact. Their scientific training should enable them to record the actions of the nonhumans and recognize and report their own reactions with a minimum of error. The anthropologist and psychiatrist will evaluate the reports in the light of each individual's personality to prevent serious misunderstandings, give a workable background for decision-making, and keep the leaders aware of crew attitudes. How we feel about a subject will eventually show itself in our actions and had best be dealt with before it is reinforced by misunderstanding or incubated by repression.

On earth, governments will need information to aid them in preparing their populations. Although the team members will not be a cross section of humanity, from their reactions the anthropologist should be able to predict which features of the nonhumans and nonhuman cultures contacted would cause strong positive or negative response on earth, and should then be able to make recommendations to leaders and administrators. On their return the expedition members themselves will be asked to write and speak about their experiences, and the attitudes that they develop during contact will profoundly affect public opinion.

Thus the two areas most amenable to modification and control will be the human element in the contact and the human element back on earth. In space the contact group will decide whether or not to contact a particular culture and how to go about it. On earth the public can accept these beings and support further contact and exploration or reject the very idea of nonhuman intelligence and withdraw from space.

Exploration will proceed in a series of ordered steps. Since man is not likely to find intelligent life at the first star, he will visit as many as time and supplies allow. The trip will include transportation from earth to star, movement between stars and the return home. At each star the team will investigate the system and locate any planets. If they find a planet they will evaluate its habitability, physical resources, and life forms. If any signs of intelligent life are discovered the survey team will have to decide whether to withdraw or attempt contact. This will involve careful observation from a distance to acquire information before actually making contact. At first, in order to gain language skills and social understanding, contact will be limited to individuals in small groups. In this way we can increase the chance of success at the official, formal meeting with extraterrestrial leaders to arrange recognition and continued contact.

There could be no crew/scientist distinction or hierarchy on the spaceship. Like our astronauts, every member of this community will be a practical scientist capable of original thinking. At various stages of exploration the leader/worker relationship would depend on the goals at that time, with a definite role for each member of the team. Except for the captain, each person would have both leader and worker areas of duty. An individual's leader specialty might be biology, with wide general knowledge and experience that would enable him to use the ship library for detailed botanical and zoological research. As a worker or team member, he could work under the engineer in such life-support areas as hydroponics and animal care. All team members would have a share of housekeeping duties.

Flexibility would be essential for survival and efficiency. If one team member is lost or absent, other members must be able to carry on his duties. It is normal human behavior to establish territory, and this must be taken into account in choosing personnel and assigning jobs. Each job must have its boss and apprentices. A situation with stress and potential danger is not the time to change human nature. Furthermore, by working in several areas, each member's store of experience and knowledge would be available for problem-solving in areas besides his own specialty. Often one person's not quite workable idea will stimulate another to develop a better one.

The exploration party would share information and ideas in two ways: progress meetings at regular intervals chaired by the captain, and problem meetings initiated and chaired by the scientist faced with a specific problem. Progress meetings would include formal reports from each department, while problem meetings would cope with the problem area in informal discussions.

The extraterrestrial journey would be made up of five different stages: the trip, the discovery, the exploration, the survey, and finally the actual contact. All job assignments would be defined in terms

of the stages of the journey, since not all members of
the expedition team would be needed at every stage.
Below is a list of job descriptions and a personnel
and organization chart.

GENERAL JOB DESCRIPTIONS:

Captain	In charge at all stages.
Navigator	Second in command, astronomer (first mate).
	In charge of discovery team once star is reached.
	Member of exploration team (physical sciences).
Communications	Third in command (second mate).
	In charge of communications, computer and records.
	Member of survey team XTI.
Engineer	Fourth in command (third mate).
	In charge of life-support systems, transportation, supplies, construction.
Geologist	In charge of physical exploration team.
	Assist engineer during trip.
	Assist navigator on discovery team.
Biologist	In charge of life exploration team.
	Assist engineer during trip and discovery periods.
	On survey team (physical anthropology areas).
Anthropologist	In charge of survey team XTI (cultural and physical anthropology), alert at all stages for signs of XTI.
	On exploration team (life sciences).
	On contact team to advise ambassador/captain.
	Assist engineer during trip and discovery.
Doctor	Medical/psychiatrist in charge of health and welfare of team members.
	On survey and contact teams.

Given (Personnel and Organization Chart for Extraterrestrial Human Community)

PERSONNEL	STAGE OF JOURNEY				
	1) Trip	2) Discovery	3) Exploration	4) Survey XTI	5) Contact
Title	Captain	Captain	Manager, coordinator, captain		Ambassador
Captain	Captain	*In charge of team (physical)*	on team (physical)		
Navigator	First mate	Astronomer			
Communications	2nd mate, communications, computer, librarian (information and recreation), records				
Engineer	3rd mate, transportation (space and local), life-support systems, equipment, etc.				
Geologist	Assist engineer	on team (physical)	*In charge of physical team*	on team (physical anthro.)	
Biologist	Assist engineer	on team (life)	*In charge of life team*	on team	on team
Anthropologist	Assist engineer	on team (XTI)	on team (life, XTI)	*In charge of team (cultural anthro.)*	on team
Doctor	Medical/psychiatric health and welfare of crew				on team

Each member should have some experience of the others' jobs.

During the trip the anthropologist would study and record the interactions of the members of this small, specialized community and develop an understanding of their personalities. He would also do his share of life-support and household duties under the engineer. As a member of the discovery and exploration teams, he would be particularly alert for signs of XTI. Since the intelligence level of the extraterrestrial culture could be hard to detect and define, the anthropologist would have to instruct the exploration members on how to recognize signs of intelligence. The geologist might notice surface features that could not be explained by natural processes, such as canals and irrigation systems that seem to have been designed. This could provide a major clue to the understanding of XTI.

The survey team's only concern would be nonhuman extraterrestrial intelligence. The anthropologist heading this team would be a xenothropologist, a specialist in nonhuman intelligent beings. He would be initiating new vocabulary and following new directions of thought and action that would affect the future growth of this science. Team members under him would be: a biologist for physical information; a doctor of psychiatry and general medicine who could supply information and techniques and who could observe and anticipate the effects of contact on the human community; and a communications expert who would provide equipment, skills, information and observation records.

The contact team would be in charge of meeting officially with governmental leaders. The anthropologist would advise the captain-ambassador on which groups to choose for this contact. He would familiarize himself with the language and social form of a particular culture so that he could warn the contact team of any potential problem areas in the contact interaction.

We will assume that any planet that has life will have gravity, atmosphere, solar radiation, water, and land, that it will have certain ecological characteristics in common with earth, and that we will find life forms that are similar to those on earth. Life will be anchored or mobile, wet or dry. An extraterrestrial animal living in water will adapt to the same shaping forces as our water animals. A jellyfish, for example, could not live on land because of the pull of gravity—although it might adapt to air if it could secrete a lighter-than-air gas to fill its body.

Starting from these assumptions I have developed a Xeno-mammal to represent the intelligent life met by our explorers. This mammal strongly resembles the Australian marsupial, specifically the large, upright kangaroo. We will therefore call them Marsupians. They stand upright and have an elongated pelvis. They have arms, hands, and two pairs of legs. A brood pouch on the back at the shoulders provides care for the immature young. Their spine ends in a short, heavily padded tail which balances their upper torso when they move and provides a resting place when they work. Like earth mammals, the skeleton is internal, rigid and jointed. The skin is covered by a heavy, but flexible coat of opalescent scales which can be raised and lowered to control body heat, thus eliminating the need for clothing. They communicate emotion by changing color. The voice is used only to convey information. The Marsupian has a large skull, a small muzzle and small, upright ears. He has a strong sense of taste, a highly developed brain and three-dimensional vision.

The fictional development of the Marsupian method of reproduction is based on two assumptions.

One is that the adaptation of organisms to their environment can best be achieved by an exchange of genes. The other is that since the male/female role is not rigid on our world it need not be limiting here as long as genes were exchanged. On earth there are animals that change sexes—oysters and swordtail fish; males that brood the young—seahorses; and even pairs where both parents produce food—fish and pigeons.

There are four body types in Marsupian society: two sexes and two nonsexes. Each sexual Marsupian exists as part of a pair of bonded individuals who exchange germ plasm. Each member of the pair bears an immature infant which is placed in the brood pouch. The two infants raised by a pair are bonded for life to the adult Marsupians, who have control of the inheritance of the next generation. This control was an important factor in the rapid development of intelligence and culture. It works in this way: Two Marsupian pairs decide that their genes would complement or strengthen each other and the births of the infants are coordinated. Immediately after birth, two of the four infants are exchanged and each pair raises two children who are desirable marriage partners. Voluntary genetic selection is in this way a built-in part of the breeding pattern.

In each sexual pair, one member is the keeper and the other a giver. The keeper retains its infant, who will be the keeper of the next pair. These sexual Marsupians produce a milklike food in the brood pouch which contains hormones. One important use is as food for the infants. The sex of the child is determined in the brood pouch during the first few days after birth by the hormone composition of the food. The children are exchanged at regular intervals after this so that they are bonded to both parents. As the pair grow and leave the pouch they begin to produce a pre-milk that they exchange. This completes the bonding. As they develop the hormone content changes to stimulate sexual maturity. The only time that this hormone-food is exchanged outside of the pair is before infants are exchanged. The pair-bond is the earth equivalent of a marriage and the rituals that accompany the choice of an exchange pair and their sharing of hormone food parallel our marriage ceremony.

In infancy, school and work the sexual Marsupians are treated as one complementary pair, a necessity since normal growth, maturity and childbearing depend on a regular exchange of hormone food. They are about five and a half feet tall and humans are unable to discern the difference between the keeper and the giver member of a pair. In fact, equality of sex is built into their life style because both members of the pair have the same duties.

The Marsupian leaders are nonsex individuals who have been raised as a singleton by two parents. They are larger than the sexual Marsupians—in fact, over six feet—because they have received food for two during early development, but they fail to develop sexually because they had no partner with whom to exchange hormone food. Leaders receive early, intense schooling, more advanced than the pair-bonded children's education. Higher education may be needed for jobs that demand independence, mobility and constancy, such as scientist, police officer, business manager, teacher, government official, religious leader and explorer. Leader status does not always mean a high intelligence or strong ad-

ministrative ability, but leaders are free from reproductive responsibilities and can fill jobs that are
more demanding.

The workers are nonsex Marsupians who develop
when one parent raises two children. They are small
(about five feet tall) and imprinted to communal
living by having shared food and living space during
early growth. Since they are brought up by only one
parent, they do not have the hormone stimulus to
develop sexually, and after early schooling they are
apprenticed to a guild group and receive training in
that speciality. The guilds train workers for blue-
collar jobs as well as more scientific and technical
activities.

The social stability in this culture results from
the close family ties that each group has with other
groups. It is also helpful that each group feels that it
has advantages over the others. Pairs are proud of
reproduction, but leaders and workers value their
freedom.

In meeting a Marsupian for the first time, an
explorer from earth would be most intrigued by the
Marsupian's six extremities, the brood pouch, and
the tail which is used in place of a chair. In an
attempt to ease the tension of contact, the anthropologist might suggest that the humans conform to
the peculiarities of the Marsupian body shape. For
example, the anthropologist might have the engineer
construct a chair that resembled an English walking
stick that the humans would carry and lean upon
much in the same way that the Marsupians use their
tails. To the Marsupians, a human shape would be
strangely proportioned. In their eyes we would be
too tall for our weight and deformed, lacking a pair
of legs and a tail. A man-made walking stick would

make us look more normal, and this in turn would help communication.

Since most of the explorers from earth would be under six feet tall, they would not be considered by the over-six-foot-tall leader-administrator Marsupians to be equal in rank. Thus, it would be important to have a tall member in the human contact group to partake in the first official meetings.

There are several Marsupian cultural taboos that we would have to respect. The crew members would have to restrain themselves from peering at brood pouches, for instance, since in Marsupian culture this is a very rude thing to do. Marsupians also have a strong taboo concerning the public exchange of hormone food which extends to touching hands and putting the hands to the face. Therefore, when they work together they turn blue. This color change inhibits food production temporarily in sexual Marsupians and thus allows them freedom to touch and handle objects together. Since the explorers could not change color, the Marsupians would feel hampered in their communication with them unless they could find a plausible substitute. A possible solution could be to use sheer, brightly colored silk scarves as body coverings. The walking stick and the silk scarves may seem like trivialities, but learning another organism's body signals and becoming comfortable in each other's presence are very important factors for insuring successful contact.

There are certain parallels between the social institutions of the Marsupians and those on earth. The institution of marriage is similar to some Eastern cultures in that marriages are arranged before the children are born, although the reasons are different. The Marsupian reproductive cycle dictates when the marriage arrangements are made and their knowledge of genetics influences the selection of mates.

Marsupia has a number of local governments which function for the same reasons and with as much variety as we find on earth. The voting class is comprised of the pair-bonds who elect leaders to positions of authority that range from church elder to prime minister. Leaders and workers do not vote. The theory is that self-interest might motivate a leader's ballot decision and workers have their own subgovernment in each guild. Leaders and workers choose members of their own groups to represent them in the lawmaking and judicial divisions of government.

War is regarded with revulsion. Any action causing injury or death to the individual, especially the pair-bond, is taboo, and guilt feelings and punishment are severe. There is not a great deal of crime, and certainly the fact that this is considered an undesirable heritable trait has its effect.

Marsupian culture is not as industrialized as earth's, but the Marsupians would be advanced enough in science and technology to be able to understand and adopt our knowledge. We would have to be cautious about introducing technology to a more primitive society for their protection. If guns were brought to defend us against large animal life, for instance, they would have to be locked up until their full danger was clearly understood by the nonhumans.

The Marsupian religion is science. The science of genetics was first developed from their church's extensive genealogical records, which were kept to assist in partner-pair choice. Their religion reflects their ability to control inheritance, and their ultimate goal is to learn the rules of the universe so that they can predict and control their environment. They are a much younger race than humanity and are still developing.

In some ways the Marsupians have developed a more workable culture than ours; yet they would probably feel the same way about earth's developments in science and technology. We must remember that these are two different cultures which have evolved from very different physical and psychological needs. The Marsupian physical structure dictates many of the social customs that we would find so desirable. We want peace, health and stable family life, but would we be willing to use their breeding methods to attain these goals? They would realize that we are more advanced in the hard sciences, but would they be willing to have a war and armed neutrality to stimulate the development of rockets and atomic power? Each culture develops in answer to the pressures of its surroundings and each has something of value to offer the other.

For contact to be worth the effort, time and money involved, there must be continued trading of information and material. For example, in our contact with the Marsupian culture, the following exchanges might occur. If they generate and use biological electricity but have not learned to store it chemically in small containers, they could receive batteries and knowledge from us and we could learn about the biological storage of energy from them. Or in areas where their biological sciences have taken precedence over physical sciences, they could acquire our specialized equipment and we could learn about the biological production of useful quantities of light.

The creation of a Marsupian culture has provided an example of how human and nonhuman cultures could affect each other. Anthropology must use its skills and attitudes to meet the challenge of contact, not by judging cultures by what is considered normal human behavior, but by recognizing and accepting cultural differences. By anticipating problems of adjustment, anthropologists can act as buffers and advisers to insure that contact with an alien population occurs with a minimum of disruption. With controlled conditions and intelligent leadership, the contact group will have the greatest chance for a fruitful exchange of information to occur. We and they will be free to react according to our respective society's normal way of dealing with new and unusual problems, modified by the knowledge of the other's limited awareness and intentions of good will.

Note

1. Stephen H. Dole and Isaac Asimov, *Planets for Man* (New York: Random House, 1964).

8.
Earth Colonization of the Moon and the Effects of Alien Contact

KIM ARTHUR MAYYASI

Mankind's first outpost in space is likely to be on the moon. What kind of cultural patterns will a moon community establish after the preliminary investigating teams have left? The members of this community will be highly qualified and capable of assuming many responsibilities. They will be above average—that is, more highly disciplined and intelligent than the norm. They will probably include students, scientists, technicians and others associated with moon-related fields of interest or study.

The gravity of the moon is only one-sixth that of the earth. This means that human muscles, including the heart, would tend to atrophy, and the sudden increase of earth's gravity after an extended stay on the moon would be fatal. Presumably people would be unwilling to inhabit a place as unattractive as the moon unless they knew that they could return to their mother planet. For this reason, as well as for obvious psychological ones, the moon society would have to be a rotating one. People would be able to stay for only a few years. But rather than attempting a complete turnover, where the entire lunar community would be replaced by an equal number of immigrants from earth, a small group of earthlings would replace a group of moon inhabitants from diverse areas of lunar life at given time intervals.

Under these conditions the community would have to have a very limited government, if any. Because the majority of the decisions would not be policy making but immediate, urgent decisions, the "government" should consist of a base commander who would be in charge of various branches.

Policy-based decisions would be made on earth in collaboration with the moon inhabitants. Since the lunar community probably would not be the world cooperative venture one would hope it to be, there would be two separate communities, each representing the two major space powers. Whether the two would join on the moon would depend on what happens on earth. The sociological situation would be similar to that which exists on ocean oil derricks drilling many miles offshore. The crews work in a harsh environment twenty-four hours a day. They are on for a period of time and off onshore for even longer. There is one "boss," a foreman, and there are several other men in charge of the drilling operations. The men are concerned only with accomplishing their work quota and getting back to shore. Any policy-making decisions are made on land and implemented in the next shift, or radioed aboard.

The rotating society would not be conducive to the creation of new cultural patterns, since the people would not sever their ties with their mother planet but rather do everything they could to bring earth ways to the moon. However, the rotations of population and the fact that the potentially harsh environment would require sophisticated machinery and equipment to keep people alive and productive would cause disruptions in the moon-synthesized earth culture. Although the lack of a strong government would create an unrestricted atmosphere, the regulations needed for survival would be enforced by all who inhabit the living structures, for the mistake of one could jeopardize the lives of many. The unconscious awareness of this fact would induce a sense of responsibility much greater than any formal governmental policy could hope to accomplish. And thus the society would be well structured and, more important, resistant to change. At the same time, however, the earth culture would probably be changing at a rapid pace, leading to further distortion of the earth's cultural patterns in the moon colony.

An interesting and helpful parallel can be drawn between a newcomer to the moon and a child going to camp for the first time. When the child arrives, his fears and anxieties climax. He makes an extra effort to avoid breaking rules, for fear of bringing an imagined hostile world down on his head. The next step is a change in his values. Overnight he becomes attached to shiny stones or a turtle to which he formerly would not have given a second thought. He starts running around without a shirt or shoes. A campfire takes on great dimensions. Handicrafts such as carving or whittling become his common pastimes. This is a reversion triggered by environmental stimuli. On a moon base, the environment would be

artificial rather than natural, and the people, of
course, would be relatively sophisticated adults
rather than children. But the comparison still holds.

The cultural patterns that will develop on the
moon will be affected to some extent by the newcomer's psychological state when he or she first lands
on the moon. Let us conceive of the following situation. Although the moon inhabitant-to-be has
undergone a training program, he is overwhelmed
by the flight to the moon. The first experience with
zero-gravity is strange but not unpleasant. To watch
earth constantly diminish in size creates a feeling of
insecurity that increases as the cold, harsh moon
grows larger. All anxieties and fears climax at the
landing. Upon arriving he finds that the living
quarters are merely functional and the food is edible,
but not delicious. The only outstanding feature of
the moon base is the scientific equipment.

The first week is an introduction to the base and
new friendships. The need for security is strong and
the newcomer to the moon tries to blend into the
established society, just as the new child at camp
anxiously tries to please the counselors and avoid
breaking the rules. This psychological dependency
on the existing society is a common psychological
state accompanying any person wishing to join a
specific community. What is important in this
unique case is that the primary method by which
cultural patterns on the moon would undergo a
transformation parallel to those occurring on earth
would be through the new members. If they were
absorbed into the moon society and its culture, they
would be less apt to initiate the cultural changes
being relayed from earth. These cultural changes
might penetrate the lunar society when more people
of the same earth-time period arrive, but more likely,

the culture would not change in direct response to earth's cultural changes, but in its own way. The people on the lunar base would be living in such a psychologically structured society and in such an artificial physical environment that change would probably come from within, initiated by the lunar inhabitants. It would be mandatory for them to adapt psychologically and sociologically to their environment in order to maintain their mental and physical health.

The social fabric of the society would be unique. Division of labor by sex would be almost nonexistent. In the early stages of the community, each individual's work load would be divided. For example, the electrical technician working with the radio telescope would also be in charge of major repairs on the communication equipment in the radio station. Once the community grows in size and begins to have a cultural pattern of its own, work becomes more specialized. Because lunar community members would be working in an exacting environment, specialization might extend beyond what exists in earth cultures.

An institution, in the sociological sense, can be defined as a group of persons who perform common activities regularly enough to develop a pattern of behavior, their own rules, and an esprit de corps. In a moon culture of increased specialization, separate and distinct institutional groups would be formed. For example, the men and women supplying fresh food, a luxury in such a place as the moon, would be working in vast underground caverns with acres of hydroponic crops. The lunar environment would force this new breed of farmer to reside in living quarters near their crops, away from the mainstream of lunar life. Clothing and other aspects of materialistic culture would be tailored to these farmers' tasks. Under these conditions, the strength of institutions and kinship groups would increase. A unique consequence of this is the institutional-kinship effect, where kinship forms of social organization would increase in direct proportion to the institutional effect. Certain extreme prejudices could be formed. For example, a newly arrived scientist working with the radio telescope may be convinced by his colleagues that the lunar inhabitants working with the hydroponic crops are nothing but "dumb farmers." The "dumb farmers" may indoctrinate the newcomers by making them believe that the scientists are snobs. So the "snobs" and the "dumb farmers" don't associate with each other. But one might think that such "immature" interaction could not happen between two groups of intelligent people. Perhaps this would be true in a community of select people living in a natural environment. However, it again must be emphasized that the conditions in which these people live induce a psychological state where rationalization of certain sociological happenings is totally different than that of earth cultures. In this case, the sense of individualism, which a human must feel he possesses, is lost. He obtains this lost feeling by establishing himself in a unique peer group, separating himself psychologically from the remainder of the moon base. It is interesting to note that the newcomer's primary psychological state changes from one extreme—acceptance—to the opposite extreme—semi-individualism.

The economy of the lunar community would be very simple. There would probably be some manufacturing, trade and tourism. Property would be owned by the entire community and inner community economics would be limited to the purchase of personal items.

Religious practices would probably increase. Living surrounded by an outer environment that could freeze or boil an individual's blood might tend to increase one's desire to be on better terms with his Maker. Religion is also a vital factor in dealing with the sense of alienation shared by the lunar inhabitants.

Earlier I spoke of how "cultural messages" received from the earth could be lost. The initiation of cultural changes on the moon could not occur until the inhabitants bringing identical "messages" from earth were in the majority, or until the change were accepted by a majority of the inhabitants present. However, not all of the cultural changes being relayed from earth would be initiated. Because the community is interwoven to a greater extent than earth communities, the initiation of a change in one area of collective behavior could create a distortion of the moon-synthesized earth culture beyond what has already been changed due to the artificial environment. Suppose a new religion were formed on earth resembling some type of science cult. A new scientist in the moon colony is a member of this cult. If this religion is accepted by a majority of the scientists, their scientific procedures, living quarters and life style will change. However, the religious practices and their effect upon the life style within the rest of the community will not change. The institutionalism of the scientists will increase. The institutional-kinship effect dictates that the same will happen to kinship. Indeed, the friction between scientists and nonscientists could become so great that work on the lunar base would be hindered. The science cult, which might have started as a "fad" on earth, could thus snowball into gigantic proportions on the lunar base. The ground controllers, realizing the problem, might then forbid religious practices altogether until the issues were resolved.

In general, the cultural distortions that occur on the moon will be due to a sudden and drastic cultural change in a group of people on earth, with little time for adjustment. Throughout its history mankind has been called upon to adapt to harsh conditions of both natural and man-made origin. These adaptations have been tested by other trials and further changed until the resulting society could survive any ordeal. This process is similar to Darwin's theory of natural selection in dealing with the evolution of animal life. But despite this adaptation process, the effect of nonhuman cultural contact with the inhabitants of the moon base would be tremendous and could be very destructive.

The earth as of now does not possess the technology to colonize beyond the moon. Thus, a confrontation between a nonlunar alien culture and the lunar community would most likely occur on the moon and the alien culture would be vastly more technologically advanced than man. The primary contact situation would be between the moon inhabitants and the alien explorers, who would be the counterparts of the earth astronauts. Thus it can be anticipated that these aliens would be ideal physical specimens and would not be characteristic of the entire alien culture, but a pseudo-militaristic part of it.

The primary contact situation should be examined fully, for the resulting impact would create a cultural shock which would play a key role in the response of the moon and earth cultures. Consider the following scenario. An alien space vessel would land on the moon's surface. An investigating team would meet the aliens. Both parties would be in spacesuits due to the vacuum and extreme temperatures. Communication would be through hand signals unless the aliens had obtained a workable knowledge of the language by monitoring radio frequencies or other means. The latter is most probable, for an intelligent race would not risk contact without a method of accurate communication. Arrangements for diplomatic exchanges would be made when the news is relayed to earth.

Following the initial nonhuman contact, even though there would be no evidence that the alien culture was malevolent, people would be swept into a state of fear. If ever there was an event that could unite the world, this would be it. The last time the world was shown that it was not the center of the universe was during the age of Copernicus. But man has always believed that even though he was not geographically in the center of the universe, he was central in essence. This belief would be shattered by this encounter. In addition, this effect would be amplified by the aliens' greater technological and sociological advancement. The modern world is proud of its achievements. To have our ultimate human endeavor, the moon community, visited by a spaceship without warning, as if on a Sunday stroll, would be an unintentional insult. Also, this would prove that the mathematicians might be correct in estimating that there are approximately one million technical civilizations in our galaxy alone, or roughly one out of every hundred thousand stars.[1] Putting the earth in its proper perspective would indeed be sobering. The effect it would have upon the moon and earth cultures would be both positive and negative.

The aliens' visit to the moon would be logical, because the moon colony could act as an intermediary between earth and the alien culture. The visit would be a psychological buffer, for, were the aliens to visit earth unannounced, panic would heighten to all-out chaos. Also, the moon inhabitants are more scientific and disciplined in their thinking, making them less likely to report inaccurate news or commit rash acts. (It is probable that the aliens have been observing the earth for a considerable length of time before contact and have witnessed the many inconsistent ways that man exists with himself.)

After the diplomatic formalities, it would be decided that a system of trading should open between the two cultures, with the moon as the go-between. Some type of alien structure and housing facilities for the aliens manning the port would be erected next to the lunar community in order to facilitate the exchanges.

One of the important factors that has not been considered is the physiology of the aliens. The appearance of the aliens would play an important role in determining the initial psychological impact of the aliens upon mankind. The first question which would be asked by those just learning about the alien contact will be, "Really! Why, what do they look like?" If the response was, "Why, just like us. They're humanoid," the reaction would be considerably more favorable than if the answer was, "Oh, something like a cross between an alligator and a boa constrictor with arms." Unfortunately, children are raised to a point where even as adults they are repulsed by reptiles. This would make the aliens appear malevolent to the majority of the human race.

The next important aspect of the alien's physiology is the atmosphere and gravity that he would need for survival. If the alien race could not tolerate normal earth conditions, then we would feel psychologically distant from them. If the aliens could, it would present an unconscious threat to know that the earth could make a good home planet for another race in our galaxy. Let us assume that the aliens can exist in an earth environment and will somewhat resemble a human, at least a biped. The cultural shock after the initial contact with the aliens would still induce a general numbness. The lunar culture, reflecting earth's culture, would now be influenced by a superior race and would try to equal it by imitation. If the lunar community were to emulate the alien society, the results would be detrimental to all, for the changes would be the lunar community's direct response to the alien's higher level of progress, although the lunar community's sociological and psychological structures would not be ready for such fantastic progress. This would result in permanent damage to the lunar system. The type of distortions which would occur cannot be hypothesized without knowing more facts. Since the template for the lunar community's social evolution would be the trading of goods and services, the chain reaction of cultural changes would begin with the economic and trade aspects of the community and then branch out. This effect would be more intense on the moon than on earth, due to the increased specialization and the close-knit, complex web of interwoven functions of community life.

Notes

1. Carl Sagan, ed., *Communication with Extraterrestrial Intelligence* (Boston: MIT Press, 1973) .

Afterword: The Inception of Extraterrestrial Anthropology

SOL TAX

Editor's Note: Sol Tax, professor of anthropology at the University of Chicago and Director of the Center for the Study of Man at the Smithsonian Institution, has initiated numerous innovative movements in theoretical and applied anthropology and is in a unique position to assess the significance of the inception of extraterrestrial anthropology. He is the father of a new clinical "Action Anthropology," which helps communities in trouble to discover and thus alleviate the underlying causes of their particular crisis. In 1960 he launched a new type of journal, *Current Anthropology,* that uses a discussion and dialogue format and publishes articles from all over the world. As president of the Ninth International Congress of Anthropological and Ethnological Sciences in 1973 he applied a new nonhierarchical principle, learned from North American Indian tribes, to the organization of the heterogeneous congress of participants from all continents. Sol Tax has received world-wide recognition for his scientific and organizational achievements.

Anthropologists have been studying the human species in its origins and its variations. But there are no fossils in the sky, and we do not yet have evidence of life, society and culture outside the earth. Without hard data, we only manifest our own imagination and yearning. Myths and dreams are objects of study by anthropologists. Therefore to entertain myths and dreams without hard data is to become ourselves objects of study. Science fiction? yes; anthropology? no.

But anthropologists *do* understand that on the deeper level the conceptual structures underlying anthropological theories are based on our *perception* of real situations, and that the recent radical change in our species' and our planet's situation must lead us to new conceptions. I remember an incident in my childhood: a schoolboy was dozing in class and woke up with a start to ask the teacher excitedly, "What was that you said?" "I said that the world will come to an end in two billion years." The boy, much relieved, said, "Oh, I thought you said two million." A world war later, the prediction of two billion suddenly collapsed to a two-digit number. It became a different world, to which even anthropology would have to respond. The doubtful future became a major human problem. The planet earth became terrifyingly small and fragile. Social immortality—posterity—appeared to be a delusion.

Scientists, including anthropologists, felt the change first. The intellectual readjustment and re-thinking began not for some unknown species populating an unknown planet, but for the species of which we are members.

Anthropology was born in the industrializing world in an era of confident progress at home and colonial domination abroad. As our knowledge of non-Western cultures deepened, it began to free us and the world from ethnocentrism, which had become the most dangerous trap of the human species. The view that human populations are equally capable, and their differing cultural heritages complementary and valuable, unmasked the moral pretensions that had been used to justify the colonial system, and hastened its end. But far from being eliminated in the world, ethnocentrism and racism are still our most dangerous traps, critically handicapping us each time we deal in a uniform manner with social problems in a heterogeneous city or nation, or advocate a solution to the world crises by promoting our own notion of progress, life style, form of technology or culture. Our planetary life expectancy may well depend on the speed with which we can discard these prejudices. If we survive long enough to make intercultural connections in outer space, it may appear later that our experience in curbing ethnocentrism on earth in the twentieth century saved us in the twenty-first from the disastrous mistakes that characterized our nineteenth.

The critical importance of this book for anthropology today is that it removes itself from our planet to view "human nature" as a whole. It envisions the opportunity to study the human behavior and the change or development of human cultures under extraterrestrial conditions; to test the applicability of anthropological knowledge to the design of extraterrestrial human communities; and to develop anthropological models for quite different species of sentient and intelligent beings by using, on a higher level, the comparative methods by which we have come to understand each earthly culture in contrast to others. All this would mean the beginning of an understanding of our species in relation to others. Comparison with other primate or mammalian or even insect species on earth has been useful in discovering the degree to which human behavior depends on culture. This method of comparison can be extended to include extraterrestrial beings. Only when we have comparisons with species that are cultural in nonhuman ways—some of them may be far more advanced than we—will we approach full understanding of the possibilities and limitations of human cultures. Even if we have no contact with nonhuman cultures in the immediate future, the models that we meanwhile make require that we sharpen the questions that we ask about human beings.

Imagination is much in evidence in the chapters of the present book, but so is a great deal of factual information from other disciplines which—anthropologists will now realize—is required in order to see the relative position of our species in the universe.

ABOUT THE EDITORS

MAGOROH MARUYAMA was born in Japan in 1929. He studied at the University of California at Berkeley, and the Universities of Munich, Heidelberg, Copenhagen and Lund, receiving a B.A. in mathematics and a Ph.D. in philosophy. He has been on the faculty at Berkeley, Stanford, Brandeis and Antioch, and is now professor of systems science at Portland State University. His work has appeared in more than sixty publications including "The Second Cybernetics" in *American Scientist,* 1963. He has been a consultant to the Office of Economic Opportunity, the National Bureau of Standards and the Canadian Ministry of State for Urban Affairs. In 1970 he organized the first cultural futuristics symposium at the American Anthropological Association meeting.

ARTHUR HARKINS was born in New York in 1936. He studied at Kansas University and the University of Massachusetts and received his Ph.D. at Kansas in 1968. He currently is an associate professor of education and sociology and the director of graduate studies at the University of Minnesota. His work has appeared in more than a hundred publications, and has been a consultant for business and industrial organizations. He was the organizer of the second cultural futuristics symposium in 1971, and a co-organizer of subsequent symposia with Maruyama.

V-285 PARKES, HENRY B.   *Gods and Men*
V-719 REED, JOHN   *Ten Days That Shook the World*
V-176 SCHAPIRO, LEONARD *The Government and Politics of the Soviet Union* (Revised Edition)
V-745 SCHAPIRO, LEONARD *The Communist Party of the Soviet Union*
V-375 SCHURMANN, F. and O. SCHELL (eds.) *The China Reader: Imperial China*, I
V-376 SCHURMANN, F. and O. SCHELL (eds.) *The China Reader: Republican China*, II
V-377 SCHURMANN, F. and O. SCHELL (eds.) *The China Reader: Communist China*, III
V-681 SNOW, EDGAR   *Red China Today*
V-312 TANNEN   *Ten Keys to Latin America*
V-322 THOMPS   *English Working Class*
V-724 WALLAC   *War an   Eve of*
V-206 WALLE   *ence*
V-298 WATTS
V-557 WEINS   *1912-*
V-106 WINST   *Cross*
V-627 WOM
V-81   WOO   *agai*
V-486 Woo
V-545 Wo
V-495 YGI   *Co*